Theo Fett

New Contributions to R-Curves and Bridging Stresses – Applications of weight functions

New Contributions to R-Curves and Bridging Stresses – Applications of weight functions

**Schriftenreihe
des Instituts für Angewandte Materialien**

Band 3

Karlsruher Institut für Technologie (KIT)
Institut für Angewandte Materialien (IAM)

Eine Übersicht über alle bisher in dieser Schriftenreihe erschienenen Bände
finden Sie am Ende des Buches.

New Contributions to R-Curves and Bridging Stresses – Applications of weight functions

by

Theo Fett

Impressum

Karlsruher Institut für Technologie (KIT)
KIT Scientific Publishing
Straße am Forum 2
D-76131 Karlsruhe
www.ksp.kit.edu

KIT – Universität des Landes Baden-Württemberg und nationales
Forschungszentrum in der Helmholtz-Gemeinschaft

KIT Scientific Publishing 2012
Print on Demand

ISSN 2192-9963
ISBN 978-3-86644-836-0

Dedicated to

Prof. Dr. Dietrich Munz

on the occasion of his

75th Birthday

Dietrich Munz worked as professor for reliability and failure analysis in mechanical engineering at the University of Karlsruhe and director of the Institute for Materials Research II at Research Centre Karlsruhe from 1980 - 2002. He has continuously supported the ceramic research as chairman of the board of the ceramic research association Karlsruhe – Stuttgart (KKS) and as Managing Director of the Institute for Ceramics in Mechanical Engineering (IKM) at the University of Karlsruhe. This institute was originally founded as a platform institute to investigate the opportunities of using high-tech ceramics in various applications in the field mechanical engineering and to teach students a ceramic-equitable design.

Professor Munz served on numerous national and international committees related to ceramics and fracture mechanics. He was one of the editors of the conference proceedings "Fracture Mechanics of Ceramics" and organized the conference in Karlsruhe in 1995. After his retirement in 2002, he is still interested in the progress of fracture mechanics of brittle materials and is often visiting our institute to meet friends and to discuss scientific problems.

The staff of the
Institute for Applied Materials -
Ceramics in Mechanical Engineering
Karlsruhe Institute of Technology (KIT)

Preface

Crack extension in ceramic materials is governed by a material property called crack-growth resistance. Most ceramics show an increase of this quantity during crack propagation, which is commonly described by so-called R-curves. Well-known reasons for such behaviour are bridging effects between opposite crack surfaces, phase transformations around the tip of a crack, and development of micro-cracking zones. Similar to these effects it seems to be possible to describe the strengthening effect of ion-exchange layers and water diffusion zones in glass by a crack resistance curve on the nm-scale.

This booklet predominantly deals with the bridging behaviour and the discussion of the observed effects in terms of the fracture mechanics weight function procedure.

About 30 years ago, Professor Dietrich Munz to whom this booklet is dedicated introduced the author in the topics of mechanical behaviour of ceramics and application of the weight function technique. He initiated and promoted the common work with help and advice. The results were published in several papers. I would like to thank him for the always-pleasant cooperation.

In addition, I have to thank Professor Michael J. Hoffmann (IAM). He gave me the opportunity to continue my work at the Institute for Ceramics in Mechanical Engineering (now: Institute for Applied Materials / Ceramics in Mechanical Engineering, IAM).

I am grateful to Dr. Stefan Fünfschilling for many discussions on R-curves and to Gabriele Rizzi (IAM) for her help by performing Finite Element computations. Thanks also must be granted to Rainer Müller (IAM) and Michael Politzky (IKET) for their support in the field of computer application.

A special word of thanks must be given to Prof. J.J. Kruzic of the Oregon State University, Corvallis, for the permission to present some of his still unpublished results on CT-specimens.

Finally, I have to thank my colleagues Dr. P. Becher (ORNL, Oak Ridge), Dr. S.M. Wiederhorn and Dr. G. Quinn (NIST, Gaithersburg), Dr. D.B. Marshall (Rockwell Int., Thousand Oaks), Prof. T. Lube (University of Leoben), Prof. G.A. Schneider and DI H. Özcoban (TUHH, Hamburg-Harburg), Dr. M. Riva, Dr. S. Wagner, and D. Creek (IAM), Dr. J. Wippler and Prof. T. Böhlke (KIT, Karlsruhe) for valuable discussions and help.

Karlsruhe, April 2012   Theo Fett

New contributions to R-curves and bridging stresses

– Applications of weight functions –

(A) R-curves, observations

(B) Fracture mechanics basic equations

(C) R-curve evaluation for silicon nitride ceramics

(D) COD for Vickers indentation cracks

(E) Stable extension of indentation cracks

(F) Bridging relation as a material property?

(G) Mode-II R-curve effects

(H) Some other reasons for shielding and R-curve

<u>APPENDIX</u>

A1 Introduction

The content of this booklet mainly deals with the experimental and theoretical treatment of R-curves caused by crack-bridging interactions. For computations, the fracture mechanics weight function procedure will be demonstrated extensively. In this introduction section, a few applications are listed without going into the details. Here only the need of weight functions may be demonstrated.

A crack-resistance curve ("R-curve") represents the stress intensity factor that has to be applied by an external load for the onset and maintenance of crack extension. The effect of an increasing crack-resistance ("R-curve behaviour") is commonly described by a relation $K_R = f(\Delta a)$ in which K_R is the stress intensity factor necessary for crack propagation by an amount of Δa (Fig. A1.1a). This crack resistance of ceramic materials is of high interest for technical applications. It can be caused by several intrinsic effects as crack-face bridging, phase transformations, micro-cracking, etc.. Especially the mathematical treatment of crack-face bridging makes use of weight functions indispensable.

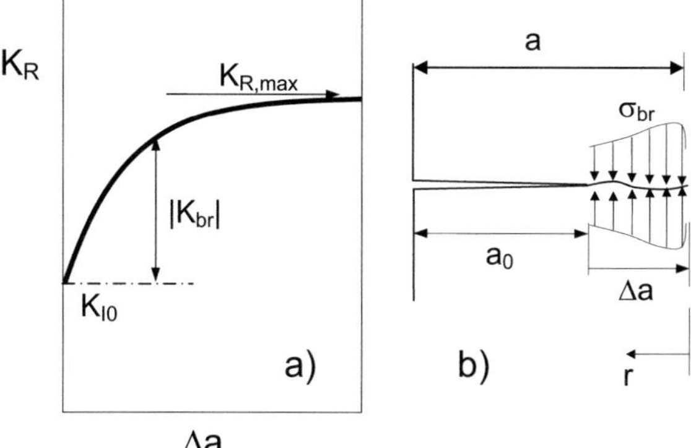

Fig. A1.1 a) Schematic of an increasing crack growth resistance curve starting from the crack-tip toughness K_{I0} and exhibiting a saturation value $K_{R,max}$, b) a crack in a ceramic material exhibiting crack surface interactions described by bridging stresses σ_{br}.

As examples for the application of the weight function method, the bridging stress intensity factor, the loading point compliance and the crack opening displacements may be addressed briefly.

In the case of a crack growing in coarse-grained ceramics, only partial crack-face separation is observed. Remaining crack-surface interactions cause so-called "bridging stresses" which act against the externally applied load (Fig. A1.1b).

Bridging stress intensity factor

The R-curve can be represented by the bridging stress intensity factor K_{br} and the starting value K_{I0}, the so-called crack-tip toughness, which is necessary for the onset of crack extension:

$$K_R = K_{I0} - K_{br} , \quad K_{br} < 0 \tag{A1.1}$$

Using the weight function representation, the bridging stress intensity factor can be represented by the distribution of bridging stresses σ_{br} acting in the wake of the crack

$$K_{br} = \int_0^{\Delta a} h(r,a)\,\sigma_{br}(r)\,dr \tag{A1.2}$$

with the fracture mechanics weight function h, the distance r from the tip, the crack extension $\Delta a = a - a_0$ and the initial crack length a_0 free of bridging.

Crack opening displacement

A popular method to determine the bridging stresses is the evaluation of crack opening displacement (COD) measured on a grown crack.

The total displacements in presence of bridging stresses result from superposition of the "bridging displacements" δ_{br} and the "applied displacements" δ_{appl} (the displacements under same load in the absence of the bridging stresses). It holds

$$\delta = \delta_{appl} + \delta_{br}$$

$$\delta_{br} = \frac{1}{E'} \int_{a-r}^{a} h(r,a')\,da' \int_0^{a'} h(r',a')\,\sigma_{br}(r')\,dr' \tag{A1.3}$$

with the plane strain modulus $E' = E/(1-\nu^2)$. The total displacement δ can be measured e.g. under the scanning electron microscope. The bridging stresses σ_{br} result by solving the integral equation (A1.3).

Compliance

For the determination of $K_R(\Delta a)$, the actual crack length, a, has to be determined. Two possibilities are widely used for this purpose:

a) Direct optical observation of the crack on a side surface of the test specimen using a microscope;

b) Measurement of the specimen compliance and computation of the actual crack length.

For metals the crack length evaluation via compliance could simply be performed since in this case free crack faces are present. In order to include the influence of the bridging interactions on the compliance, a modified procedure may be used.

As a consequence of Betti's theorem the displacements at the loading points caused by the bridging effect, $\delta_{LP,br}$, are

$$\tfrac{1}{2B} P_{appl} \delta_{LP,br} = \int_{0}^{\Delta a} \sigma_{br}(r) \delta_{appl}(r)\, dr \qquad (A1.4)$$

with the thickness B and the force P_{appl} acting at the loading points. The "applied" crack-opening displacements are

$$\delta_{appl} = \frac{1}{E'} \int_{a-r}^{a} h(r,a') K_{appl}(a')\, da' \qquad (A1.5)$$

with the applied stress intensity factor K_{appl} and the weight function h.

In the former Institute for Ceramics in Engineering (IKM) of the Karlsruhe Institute of Technology (KIT) the behaviour of R-curves due to bridging interactions has been studied in detail. This holds especially for silicon nitrides exhibiting steeply rising crack resistance. Experimental results are given in [A1.1], [A1.2], and a number of scientific papers. Theoretical work on weight functions is reported in [A1.3] and [A1.4]. In this booklet the contribution of the IKM on experimental and theoretical treatment of this type of R-curves will be compiled as resulted in the last 3 years.

<u>Section A</u> deals with some basic experiments from literature and historical aspects of R-curves.

In <u>Section B</u> the fracture mechanics basic relations are compiled which in the subsequent sections are extensively used.

The fracture mechanics evaluation of the silicon nitride qualities and the SiAlON ceramics developed at IKM is demonstrated in <u>Section C</u>. The results are compared with some commercial Si_3N_4 ceramics.

In <u>Section D</u> the R-curve behaviour of 2-dimensional surface cracks (introduced by Vickers- and Knoop- indentations) is addressed.

Indentation cracks are treated in <u>Section E</u>.

Possible effects of multiaxial and cyclic loading are addressed in <u>Sections F</u> and mode-II shielding is discussed in <u>Section G</u>.

Finally, some other reasons are compiled in <u>Section H</u> which also result in an increasing crack growth resistance.

References A1

A1.1 Fünfschilling, S., Mikrostrukturelle Einflüsse auf das R-Kurvenverhalten bei Siliciumnitrid-keramiken, Thesis, IKM 53, Universitätsverlag Karlsruhe, 2009.
A1.2 Riva, M., Entwicklung und Charakterisierung von Sialon- Keramiken und Sialon-SiC-Verbunden für den Einsatz in tribologisch hochbeanspruchten Gleitsystemen, IKM 54, KIT Scientific Publishing, Karlsruhe, 2011.
A1.3 Fett, T., Stress intensity factors, T-stresses, Weight functions, IKM 50, Universitätsverlag Karlsruhe, 2008.
A1.4 Fett, T., Stress intensity factors, T-stresses, weight functions, -Supplement Volume-, KIT Scientific Publishing Karlsruhe, Volume IKM 55 (2009).

A2 R-curves in ceramic materials

A2.1 Crack resistance

Failure of brittle materials starts from pre-existing cracks when the applied stress intensity factor, K_{appl}, reaches a critical value K_R called the crack resistance

$$K_{appl} = K_R \qquad \text{(A2.1.1)}$$

The fracture mechanics loading parameter K_{appl} is defined by

$$K_{appl} = \sigma_{appl} \sqrt{\pi a}\, F \qquad \text{(A2.1.2)}$$

where σ_{appl} is a characteristic stress value as for instance the remote tensile stress or the outer fibre tensile stress in a bending bar, a is the length of a crack and F is the so-called geometric function that depends on the type of crack, the specimen shape and the type of the load.

A2.1.1 Material property: constant K_{Ic}

For an ideally brittle material it was found that the crack resistance K_R was completely described by a constant material-specific value $K_R = K_{Ic}$ called the fracture toughness. For glasses this parameter seemed to be a true material constant in the absence of subcritical crack growth, e.g. for glass in liquid nitrogen or in vacuum (see also Sections H2 and H3).

It was very early shown by Hübner and Jillek [A2.1] that the crack resistance of polycrystalline alumina was not a constant but did depend on the crack extension Δa. The dependency $K_R = f(\Delta a)$ is known as the R-curve.

A schematic representation of an R-curve $K_R = f(\Delta a)$ is given in Fig. A1.1a in the Introduction section. The initial value of K_R at the onset of crack extension is called the crack-tip toughness K_{I0}. Very often (but not in all cases) a saturation of $K_R \rightarrow K_{R,max}$ is observed.

Consequently, the very simple failure concept of a constant failure property could no longer be used for such ceramic materials. In numerous publications of the eighties, the increase of crack resistance was reported for many ceramics. Here only a few references may be given, e.g. for coarse-grained alumina: [A2.2, A2.3], zirconia ceramics: Mg-PSZ [A2.4, A2.5, A2.6], Ce-TZP [A2.7, A2.8], and later also for silicon nitride: [A2.9, A2.10, A2.11].

For the description of the material response during crack extension, the complete R-curve was then interpreted as the true material property.

A2.1.2 Material property: $K_R(\Delta a)$

The application of R-curves for a material characterization with respect to resistance against crack extension was a rather simple and well-established procedure.

However, unfortunately, it has been shown by experimental and theoretical investigations that also the R-curve is not a unique material property. Experiments with different initial crack lengths showed that $K_R(\Delta a)$ is not really independent of test conditions but depends on the geometry of the test specimens, the initial crack depth, the type of loading (tension, bending, point forces).

Especially, R-curves for naturally small cracks are often different from those for macroscopic cracks. A survey of literature data was given by Munz [A2.12] showing the trend of lower R-curves for small natural cracks.

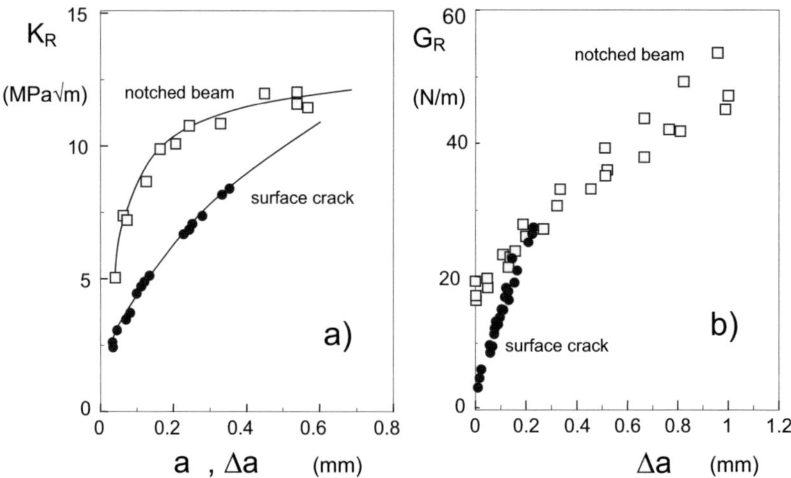

Fig. A2.1 R-curves for large and small cracks: a) MgO-doped zirconia [A2.13], b) alumina [A2.14].

Figure A2.1a shows results of macroscopic cracks (squares) for MgO-doped zirconia (Marshall and Swain [A2.13]) and Steinbrech and Schmenkel [A2.14] for coarse-grained alumina (Fig. A2.1b). The circles represent the results for surface cracks. The ordinate in Fig. A2.1b is the energy rate G_R necessary for crack propagation. It is related to K_R via

$$G_R = \frac{K_R^2}{E'} \;,\quad E' = E/(1 - v^2) \tag{A2.1.3}$$

(E'=plane strain modulus, E=Young's modulus, v=Poisson's ratio). From these data it is clearly visible that the crack resistance for the extension of small surface cracks is lower than that for large through-the-thickness cracks.

A2.1.3 Material property: bridging relation

It is actually accepted opinion that in the special case of R-curves caused by grain bridging effects, the relation between the bridging stresses and crack opening displacement, $\sigma_{br} = f(\delta)$, is the intrinsic material property which is expected to be much less influenced by test conditions.

Direct measurements of the loads transferred by the bridges were performed by Pezzotti et al. [A2.15, A2.16] and Kruzic et al.[A2.17] applying Raman spectroscopy.
Hay and White [A2.18, A2.19] developed the post-fracture tensile (PFT) test which gives the crack closure stress *vs.* the crack opening displacement relationship.

A very popular method to determine the bridging stress relation is the evaluation of crack opening displacement (COD) measurements. In the case of coarse-grained alumina often scanning electron microscopes (SEM) were used for this purpose [A2.20, A2.21] (see also [A2.22, A2.23]). This procedure will be addressed in detail in Section D.

A2.2 Reasons for the R-curve interpretation by crack-face bridging

A2.2.1 Re-notching experiments

Knehans and Steinbrech [A2.2] carried out a very pioneering experiment that gave much insight in the nature of R-curve behaviour. These authors measured the crack resistance of coarse-grained alumina with bending bars of different initial notch depths a_0/W (W=specimen width). Figure A2.2a shows the scatter bands of results for a_0/W=0.4 and 0.6 as the hatched areas. Then, a crack with a_0/W=0.4 was propagated up to $a/W\cong0.66$ under full load ($a/W\cong0.7$-0.72 after unloading). The specimens were re-notched in the range of $0.4\leq a/W\leq0.6$. The crack resistance value after re-notching was clearly smaller than without re-notching (indicated by the arrow) and matched very well to the scatter band of the tests performed for a_0/W=0.6. This was a clear indication for a localization of the strengthening mechanisms in the crack wake.

As the sources of the increasing crack resistance Knehans and Steinbrech [A2.2] proposed crack-face interactions and a possible change of the micro-crack state in the wake due to the sawing procedure. The effect of each contribution cannot be distinguished from the results of Fig. A2.2a. In principle it could be possible (but is of course highly improbable) that the whole increase of the crack resistance might be caused by the change of the micro-crack state. The fact that crack-face interactions must be present can be seen from the increasing compliance of the re-notched specimens.

A change of the micro-crack population along the notch faces results in a constant non-elastic strain contribution. Since this strain is independent of the applied load, there will not appear an effect on the specimen compliance. The occurrence of a compliance change was a rather simple proof for the crack-face interactions in the crack wake.

Figure A2.2b shows three loading-unloading cycles for a re-notching experiment carried out by Himsolt et al. [A2.24, A2.25] on a Chevron-notched bar of hot-pressed silicon carbide (HPSiC) containing 0.3 wt% Al (grain size: 1.2-4.9 µm). Whereas the grown crack shows a steep load vs. displacement curve with a hysteresis, the specimen re-notched up to 98% of the crack length is clearly more compliant. In addition, the hysteresis is strongly reduced. In the case of re-notching to 99.5%, the compliance is maximum and a hysteresis is no longer visible. Such results clearly indicate the existence of unbroken ligaments behind the crack tip.

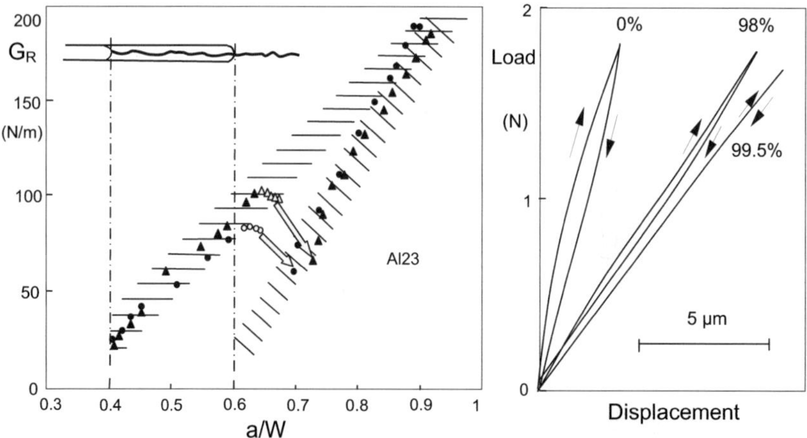

Fig. A2.2 a) Re-notching experiments on coarse-grained alumina by Knehans and Steinbrech [A2.2], b) change of compliance by re-notching of HPSiC (Himsolt et al. [A.2.24, A2.25]).

A2.2.2 In-situ observations

The existence of crack-face interactions could very early be proofed directly by the observation of crack paths under the SEM. Many images of such interaction events were published by Swanson et al. [A2.26] and Fairbanks et al. [A2.27] clearly indicating the effect for alumina where the unbroken parts in the crack wake can act as elastic and frictional bridges.

An *elastic bridging* event is shown in Fig. A2.3 for hot-pressed silicon nitride (HPSN) [A2.28]. Such a situation can result if a crack locally extends on slightly different planes simultaneously. Crack propagation on exactly the same prospective plane as for instance visible in glasses is disturbed by the microstructure of the poly-crystalline ceramics. The left part of Fig. A2.3 shows an image of a Vickers indentation crack close to the crack tip (crack coming from the left). The right illustration is a schematic of the resulting elastic bridges.

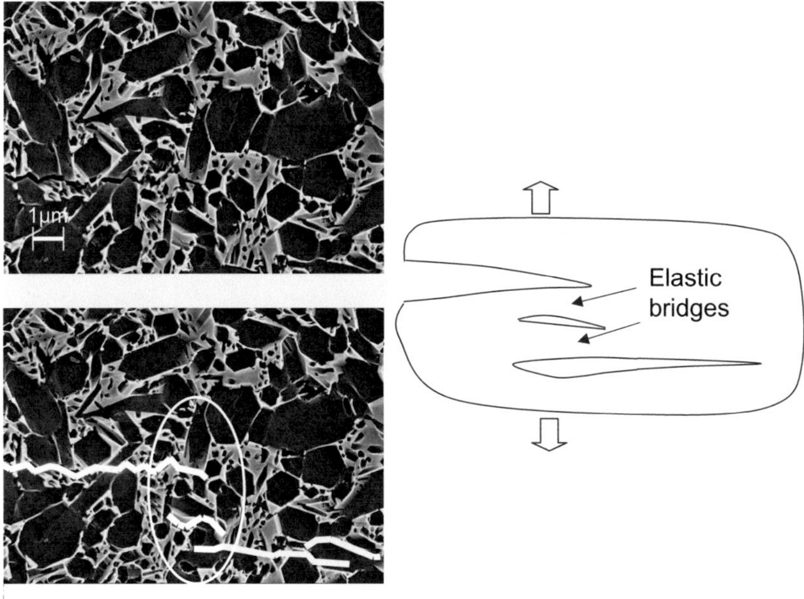

Fig. A2.3 Elastic bridging interaction for a crack in silicon nitride by Fünfschilling [A2.28].

In the *frictional bridging* model by Mai and Lawn [A2.29], tractions are transmitted between the upper and lower crack faces by friction. Large grains with the lattice orientation and the thermal expansion coefficient different from that of the surrounding matrix (assumed homogeneous and isotropic) show local residual stresses by thermal mismatch after cooling down from sintering temperature.

In Fig. A2.4, a large grain is shown, acting as a crack-bridging event. The x-component of the thermal mismatch tractions σ_{mis} is indicated. In this context, it should be mentioned that bridges show 3-dimensional crack-face interlocking with finite depth L in the order of $L \approx D$. The consequence is that mismatch stresses also act in y-direction, which is not introduced in Fig. A2.4.

During crack-face separation resulting in an increasing displacement, δ, a friction stress σ_{fr} acts which is proportional to the mismatch stress. The loads transferred by crack face interactions are localized at single grains. They can be modelled in a more homogeneous way by so-called bridging stresses σ_{br}, which average the localized interactions over a large number of grains. If σ_{mis} is the thermal mismatch stress, the bridging stress σ_{br} can be expressed by

$$\sigma_{br} \cong \mu \sigma_{mis} \tag{A2.2.1}$$

defining an *effective* friction coefficient μ.

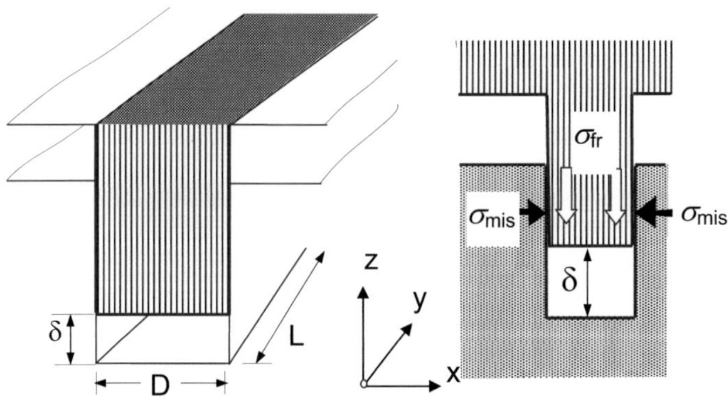

Fig. A2.4 Crack surface interactions due to a local frictional bridging event.

A2.3 Further experimental observations

A2.3.1 Influence of the initial crack length

As mentioned before the R-curve for the same type of cracks depends on the specimen and crack geometry. This could always be seen from the results of Knehans and Steinbrech [A2.2] (Fig. A2.2a). A further example is shown in Fig. A2.5.

In this diagram, two R-curves for edge-cracked bending bars with different initial crack lengths a_0 are plotted for an alumina with a grain size of 16 µm [A2.30]. The deeper initial crack shows a clearly steeper R-curve. This effect can be explained simply by the increasing regular weight function terms with increasing crack length as will be shown in Section C.

A2.3.2 Influence of temperature

Measurements of Mundry [A2.31] on a coarse-grained alumina of $d_m = 16$ µm average grain size (re-evaluated in [A2.32]) are plotted in Fig. A2.6a. From these results the bridging stress intensity factor K_{br} could be determined via the relation of eq.(A1.1) with the result shown in Fig. A2.6b. The bridging stress intensity factor K_{br} continuously decreases with increasing temperature. Since the stress intensity factor K_{br} is proportional to the bridging stresses σ_{br} as given by eq.(A1.2), the bridging stresses must decrease in the same way.

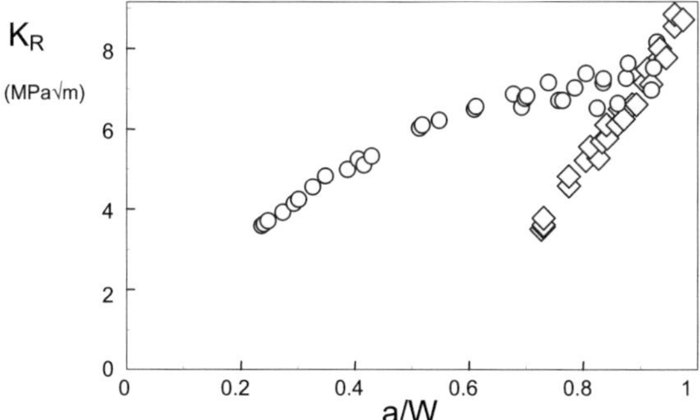

Fig. A2.5 Influence of the initial crack size on R-curves for coarse-grained alumina, from [A2.30]; (different symbols for different initial notch depths).

The same conclusion can be drawn from measurements on HPSiC [A.2.24]. In these tests carried out with Chevron specimens large crack extensions were realized. The toughness data are represented in Fig. A2.7. Already at 1000°C a toughness reduction of $\approx 20\%$ can be stated. At 1500°C, a reduction of nearly 50% is visible. The same results were obtained for specimens with straight-through notches and Knoop indentation cracks.

A simple explanation of the reduction of crack-face interactions with increasing temperature may be due to the fact that the reason for the interactions, the so-called "bridging stresses" are affected by thermal mismatch stresses. This at least holds for the friction-induced bridging events.

The thermal mismatch stresses are proportional to the anisotropy $\Delta\alpha$ of the thermal expansion coefficients in the c-axis compared to the perpendicular axes and the temperature difference between sintering temperature T_s and temperature T at which the mismatch stresses are considered. It then holds for temperature dependent expansion coefficients

$$\sigma_{mis} \propto E\int_{T}^{T_s} \Delta\alpha(T')\,dT' \qquad (A2.3.1)$$

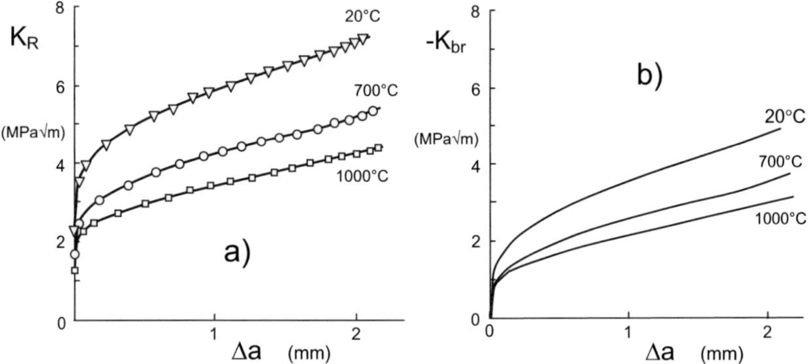

Fig. A2.6 a) Temperature influence on the R-curve of coarse-grained alumina measured by Mundry [A2.31], b) related bridging stress intensity factor.

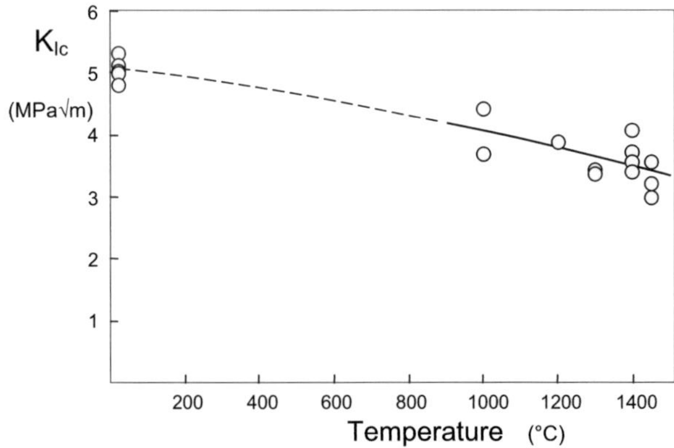

Fig. A2.7 Temperature influence on toughness for HPSiC with 0.3% Al by Himsolt et al.[A2.24, A2.25] obtained with Chevron-notched bending bars.

The temperature dependent value of $\Delta\alpha$ in the integral can be drawn out of the integral by use of the mean value theorem for integrals with the result of

$$\sigma_{mis} \propto E(T)\Delta\alpha_{eff}(T) \times (T_s - T) \qquad (A2.3.2)$$

defining an effective difference in thermal expansion coefficients $\Delta\alpha_{eff}(T)$ which depends on the temperature T at which the measurements are carried out. It is obvious from (A2.3.1) that the mismatch stresses decrease with increasing temperature T and disappear completely at $T = T_s$. Such an effect has to be expected at least for frictional bridging stresses, eq.(A2.2.1). Consequently, the contribution of frictional bridging to the R-curve must decrease with temperature.

References A2

A2.1 Hübner, H., Jillek, W., Subcritical crack extension and crack resistance in polycrystalline alumina, J. Mat. Sci. **12**(1977), 117–125.
A2.2 Knehans, R., Steinbrech, R., Memory effect of crack resistance during slow crack growth in notched Al_2O_3 bend specimens, J. Mater. Sci. Lett. **1**(1982), 327–329.
A2.3 Steinbrech, R., Knehans, R., Schaarwächter, W., Increase of crack resistance during slow crack growth in Al_2O_3 bend specimens, J. Mater. Sci. **18**(1983), 265–270.
A2.4 Swain, M.V., Hannink, R.H.J., R-curve behavior in zirconia ceramics, Adv. in Ceram. **12**(1984), 225–239.
A2.5 Swain, M.V., Rose, L.R.F., Strength limitations of transformation-toughened zirconia alloys, J. Am. Ceram. Soc. **69**(1986), 511–518.
A2.6 Heuer, A.H., Transformation toughening in ZrO_2-containing ceramics, J. Am. Ceram. Soc. **70**(1987), 689–698.
A2.7 Grathwohl, G., Liu, T.: Crack resistance and fatigue of transforming ceramics: II, CeO_2-stabilized tetragonal ZrO_2, J. Am. Ceram. Soc. **74**, 3028–3034.
A2.8 Yu, C.S., Shetty, D.K., Shaw, M.C., Marshall, D.B., Transformation zone shape effects on crack shielding in ceria-partially-stabilized zirconia (Ce-TZP)-alumina composites, J. Am. Ceram. Soc. **75**(1992), 2991–2994.
A2.9 Li, C.W., Yamanis, J., Super-tough silicon nitride with R-curve behavior, Ceram. Eng. Sci. Proc. **10**(1989), 632–645.
A2.10 Nishida, T., Hanaki, Y., Measurement of rising R-curve behavior in toughened silicon nitride by stable crack propagation in bending, J. Am. Ceram. Soc. **78**(1995), 3113–3116.
A2.11 Gilbert, C.J., Dauskardt, R.H., Ritchie, R.O., Behavior of cyclic fatigue cracks in monolithic silicon nitride, J. Am. Ceram. Soc. **78**(1995), 2291–2300.
A2.12 Munz, D., What can we learn from R-curve measurements? J. Am. Ceram. Soc. **90**(2007), 1-15.
A2.13 Marshall, D.B., Swain, M.V., Crack resistance curves in magnesia-partially-stabilized zirconia, J. Am. Ceram. Soc. **71**(1988), 399–407.
A2.14 Steinbrech, R., Schmenkel,O., Crack resistance curves for surface cracks in alumina, Comm. J. Am. Ceram. Soc. **71**(1988), C271–C273.
A2.15 Pezzotti, G., Muraki, N., Maeda, N., Satou, K., Nishida, T., In situ measurement of bridging stresses in toughened silicon nitride using Raman microprobe spectroscopy, J. Am. Ceram. Soc. **82**(1999), 1249-56
A2.16 Pezzotti, G., Ichimaru, H., Ferroni, L., Raman Microprobe Evaluation of Bridging Stresses in Highly Anisotropic Silicon Nitride, Am. Ceram. Soc. **84**(2001), 1785-90.

A2.17 Kruzic, J.J., Cannon, R.M., Ager III, J.W., Ritchie, R.O., Fatigue threshold R-curves for predicting reliability of ceramics under cyclic loading, Acta Mater. **53**(2005), 2595-2605.

A2.18 Hay, J. C., White, K. W., Grain-Bridging Mechanisms in Monolithic Alumina and Spinel, J. Am. Ceram. Soc., **76**(1993), 1849-1854.

A2.19 K. W. White and J. C. Hay, The Effect of Thermoelastic Anisotropy on the R-curve Behavior of Monolithic Alumina, J. Am. Ceram. Soc., **77**(1994), 2283-2288

A2.20 Rödel, J., Kelly, J.F., Lawn, B.R., In situ measurements of bridged crack interfaces in the scanning electron microscope, J. Am. Ceram. Soc. **73**(1990), 3313-18.

A2.21 Fett, T., Munz, D., Seidel, J., Stech, M., Rödel, J., Correlation between long and short crack R-curves in alumina using the crack opening displacement and fracture mechanical weight function approach, J. Am. Ceram. Soc. **79**(1996), 1189-96.

A2.22 Yu, C.T., Kobayashi, A.S., Fracture process zone in SiCw/Al$_2$O$_3$, Ceram. Eng. Proc., **14**(1993), 273-281.

A2.23 Fett, T., Munz, D., Yu, C.T., Kobayashi, A.S., Determination of bridging stresses in reinforced Al$_2$O$_3$, J. Am. Ceram. Soc. **77**(1994) 3267-3269.

A2.24 Himsolt, G., Bestimmung des Bruchwiderstands keramischer Werkstoffe mit Biegeproben unterschiedlicher Kerbform bei Raumtemperatur und hohen Temperaturen, PhD-Thesis, University of Karlsruhe, 1985.

A2.25 Himsolt, G., Fett, T., Keller, K., Munz, D., Fracture toughness measurements on silicon carbide, Mat.-wiss. u. Werkstofftech. **20**(1989),148-153.

A2.26 Swanson, P.L., Fairbanks, C.J., Lawn, B.R., Mai, Y.W., Hockey, B.J., Crack-interface grain bridging as a fracture resistance mechanism in ceramics: I, Experimental study of alumina, J. Am. Ceram. Soc. **70**(1987), 279-289.

A2.27 C.J. Fairbanks, B.R. Lawn, R.F. Cook and Y-W. Mai, Microstructure and the Strength of Ceramics, Fracture Mechanics of Ceramics, R.C. Bradt, A.G. Evans, D.P.H. Hasselman and F.F. Lange, eds., Plenum, New York, Vol. 8(1986), 23.

A2.28 Fünfschilling, S., Determination of R-curves from corrected load-displacement data and the calculation of the bridging stresses from the R-curves, 6th International Conference on Nitrides and Related Materials (ISNT 2009), 15.-18.03.2009, Karlsruhe.

A2.29 Mai, Y., Lawn, B.R., Crack-interface grain bridging as a fracture resistance mechanism in ceramics: II. Theoretical fracture mechanics model, J. Am. Ceram. Soc. **70**(1987), 289.

A2.30 Munz, D., Fett, T. (1999), CERAMICS, Failure, Material Selection, Design, Springer-Verlag, Heidelberg.

A2.31 Mundry, G., Gefüge, Bruchmodus und Rißwiderstandsverhalten von Aluminiumoxidkeramik unterschiedlicher Silizium- und Magnesiumoxidgehalte, Fortschrittsber. VDI, Series 18, No. 95.

A2.32 T. Fett, D. Munz, G. Thun, H.A. Bahr, Evaluation of bridging parameters in Al$_2$O$_3$ from R-curves by use of the fracture mechanical weight function, J. Am. Ceram. Soc. **78**(1995), 949-51.

B1 Basic equations for 1-dimensional cracks

B1.1 Near-tip stress field

The complete stress state in a cracked body can be determined from the Airy stress function Φ. This function results as the solution of $\Delta\Delta\Phi=0$. A series representation for Φ was given by Williams [B1.1]. In polar coordinates r, φ with the crack tip as the origin (see Fig. B1.1) the *symmetric part* of the Airy stress function, which is of interest for R-curve problems, reads

$$\Phi = \sum_{n=0}^{\infty} r^{n+3/2} A_n \left[\cos(n+\tfrac{3}{2})\varphi - \frac{n+\tfrac{3}{2}}{n-\tfrac{1}{2}} \cos(n-\tfrac{1}{2})\varphi \right]$$

$$+ \sum_{n=0}^{\infty} r^{n+2} B_n [\cos(n+2)\varphi - \cos n\varphi] \tag{B1.1.1}$$

where the coefficients A_n and B_n are proportional to the applied load. The geometric data are explained in Fig. B1.1.

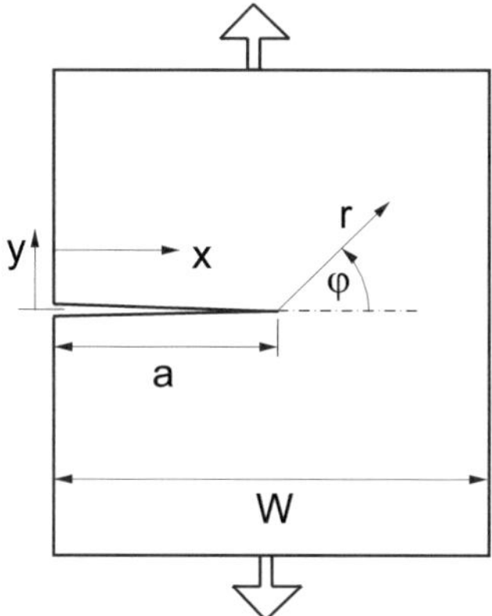

Fig. B1.1 Geometrical data of a crack in a component under tensile loading.

The tangential, radial and shear stresses result from the stress function by

$$\sigma_\varphi = \frac{\partial^2 \Phi}{\partial r^2} \ , \tag{B1.1.2a}$$

$$\sigma_r = \frac{1}{r}\frac{\partial \Phi}{\partial r} + \frac{1}{r^2}\frac{\partial^2 \Phi}{\partial \varphi^2} \qquad \text{(B1.1.2b)}$$

$$\tau_{r\varphi} = \frac{1}{r^2}\frac{\partial \Phi}{\partial \varphi} - \frac{1}{r}\frac{\partial^2 \Phi}{\partial r \partial \varphi} \qquad \text{(B1.1.2c)}$$

For the near-tip stresses the first term with $n=0$ is of interest, exclusively. The related stress components can be expressed in an x-y- coordinate system by

$$\sigma_{ij} = \frac{K_{\mathrm{I}}}{\sqrt{2\pi r}}\, f_{ij}(\varphi) + \sigma_{ij,0} \qquad \text{(B1.1.3)}$$

with the angular functions

$$f_{xx} = \cos\left(\frac{\varphi}{2}\right)\left[1 - \sin\left(\frac{\varphi}{2}\right)\sin\left(\frac{3\varphi}{2}\right)\right] \qquad \text{(B1.1.4)}$$

$$f_{yy} = \cos\left(\frac{\varphi}{2}\right)\left[1 + \sin\left(\frac{\varphi}{2}\right)\sin\left(\frac{3\varphi}{2}\right)\right] \qquad \text{(B1.1.5)}$$

$$f_{xy} = \cos\left(\frac{\varphi}{2}\right)\sin\left(\frac{\varphi}{2}\right)\cos\left(\frac{3\varphi}{2}\right) \qquad \text{(B1.1.6)}$$

and

$$\sigma_{ij,0} = \begin{pmatrix} \sigma_{xx,0} & 0 \\ 0 & 0 \end{pmatrix} = \begin{pmatrix} T & 0 \\ 0 & 0 \end{pmatrix} \qquad \text{(B1.1.7)}$$

with r and φ according to Fig. B1.1. The parameter K_{I} is the mode-I stress intensity factor and T the so-called "T-stress" term. These terms are related to the stress function coefficients by

$$K_{\mathrm{I}} = A_0\sqrt{18\pi} \quad , \quad T = -4B_0 \qquad \text{(B1.1.8)}$$

In most cases, the stress intensity factor K_{I} is expressed as

$$K_{\mathrm{I}} = \sigma^*\sqrt{\pi a}\, F\left(\tfrac{a}{W}\right) \qquad \text{(B1.1.9)}$$

where a is the crack length, W is the width of the component, and σ^* is a characteristic stress in the component, e.g. the remote tension in a tensile test or the outer fibre tensile stress in a bending bar. F is called the "geometric function", sometimes also the "shape function". It depends on the crack/component geometry as well as on the special loading.

Leevers and Radon [B1.2] proposed a dimensionless representation of T by the stress biaxiality ratio β, defined as

$$\beta = \frac{T\sqrt{\pi a}}{K_{\mathrm{I}}} \qquad (B1.1.10)$$

B1.2 Weight function procedure

The weight function procedure developed by Bückner [B1.3] simplifies the determination of stress intensity factors. If the weight function is known for a crack in a component, the stress intensity factor can be obtained by multiplying this function by the stress distribution and integrating it along the crack length. The weight function does not depend on the special stress distribution, but only on the geometry of the component.

The method is considered below for the case of an edge crack. If $\sigma_n(x)$ is the normal stress distribution in the uncracked component along the prospective crack line of an edge crack, the stress intensity factors are given by [B1.4]

$$K_{\mathrm{I}}^{(1)} = \int_0^a h_{11}(x,a)\sigma_n(x)\,dx \qquad (B1.2.1a)$$

$$K_{\mathrm{II}}^{(1)} = \int_0^a h_{21}(x,a)\sigma_n(x)\,dx \qquad (B1.2.1b)$$

These equations define the weight functions h_{11} and h_{21}. For shear tractions τ_{xy} along the crack faces it results

$$K_{\mathrm{I}}^{(2)} = \int_0^a h_{12}(x,a)\tau_{xy}(x)\,dx \qquad (B1.2.2a)$$

$$K_{\mathrm{II}}^{(2)} = \int_0^a h_{22}(x,a)\tau_{xy}(x)\,dx \qquad (B1.2.2b)$$

with the weight functions h_{12} and h_{22}. Under a combined crack-face loading, the stress intensity factors can be superimposed

$$K_{\mathrm{I}} = \int_0^a [h_{11}(x,a)\sigma_n(x) + h_{12}(x,a)\tau_{xy}(x)]\,dx \qquad (B1.2.3)$$

$$K_{\mathrm{II}} = \int_0^a [h_{21}(x,a)\sigma_n(x) + h_{22}(x,a)\tau_{xy}(x)]\,dx \qquad (B1.2.4)$$

In most practical applications and for most R-curve problems the mixed weight functions h_{12} and h_{21} disappear. In such cases we can simply write

$$K_{\mathrm{I}} = \int_0^a h_{\mathrm{I}}(x,a)\sigma_n(x)\,dx \tag{B1.2.5}$$

$$K_{\mathrm{II}} = \int_0^a h_{\mathrm{II}}(x,a)\tau(x)\,dx \tag{B1.2.6}$$

B1.3 Test specimens

Solutions for the stress intensity factor, the T-stress, and the weight functions of the R-curve-relevant crack problems considered in this book were taken from the fracture mechanics literature. Equations without reference are from [B1.4], [B1.5] and [B1.6]. In all equations it is $\alpha=a/W$.

B1.3.1 Edge-cracked bar

Very often R-curves are measured in bending tests. For a long edge-cracked bar in 4-point bending with $H/W \geq 1.5$ (Fig. B1.2) the geometric function for the stress intensity factor reads

$$F = \frac{1.1215}{(1-\alpha)^{3/2}}\left[\frac{5}{8} - \frac{5}{12}\alpha + \frac{1}{8}\alpha^2 + 5\alpha^2(1-\alpha)^6 + \frac{3}{8}\exp(-6.1342\alpha/(1-\alpha))\right] \tag{B1.3.1}$$

the biaxiality ratio

$$\beta = \frac{-0.469 + 1.2825\alpha + 0.6543\alpha^2 - 1.2415\alpha^3 + 0.07568\,\alpha^4}{\sqrt{1-\alpha}} \tag{B1.3.2}$$

and the weight function for $\alpha \leq 0.85$

$$h = \sqrt{\frac{2}{\pi a}}\,\frac{1}{\sqrt{1-x/a}(1-\alpha)^{3/2}}\left[(1-\alpha)^{3/2} + \sum_{m,n=0}^{m=4,n=2} D_{nm}(1-x/a)^{m+1}\alpha^n\right] \tag{B1.3.3}$$

The coefficients for (B1.3.3) are listed in Table B1.1.

Table B1.1 Coefficients D_{nm} for eq.(B1.3.3)

m	n=0	1	2	3	4
0	0.4980	2.4463	0.070	1.3187	-3.067
1	0.5416	-5.0806	24.3447	-32.7208	18.1214
2	-0.19277	2.55863	-12.6415	19.763	-10.9986

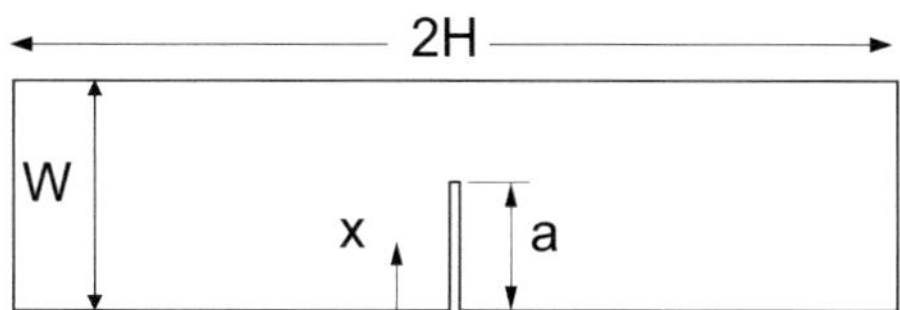

Fig. B1.2 Geometric data of an edge-notched bar.

B1.3.2 Compact tension (CT) specimen

A stress intensity factor solution for the standard CT specimens (Fig. B1.3) was proposed by Newman [B1.7]

$$K_{\mathrm{I}} = \frac{P}{B\sqrt{W}} \frac{(2+\alpha)(0.886 + 4.64\alpha - 13.32\alpha^2 + 14.72\alpha^3 - 5.6\alpha^4)}{(1-\alpha)^{3/2}} \qquad (B1.3.4)$$

The weight function solution for the CT-specimen is for $0.2 \leq \alpha \leq 0.8$ [B1.4]

$$h = \sqrt{\frac{2}{\pi a}} \frac{1}{\sqrt{1 - x/a}(1-\alpha)^{3/2}} [(1-\alpha)^{3/2} + \sum_{m,n=0}^{m=4,n=3} D_{nm}(1 - x/a)^{m+1}\alpha^n] \qquad (B1.3.5)$$

with the coefficients listed in Table B1.2

The biaxiality can be described by

$$\beta \cong \frac{0.7702 - 6.572\alpha + 26.665\alpha^2 - 43.446\alpha^3 + 29.695\alpha^4 - 6.6886\alpha^5}{\sqrt{1-\alpha}} \qquad (B1.3.6)$$

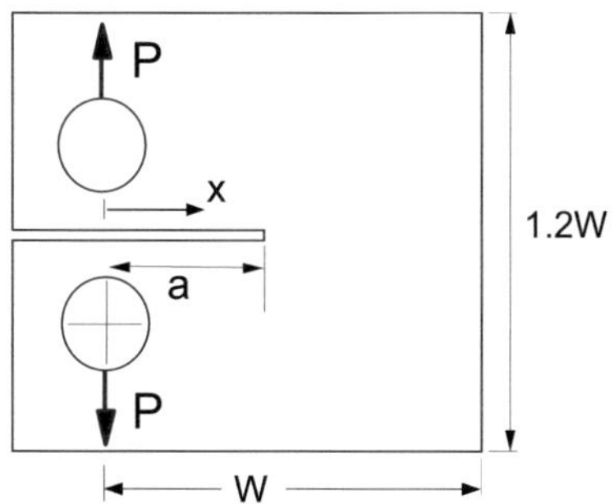

Fig. B1.3 Geometric data of a CT-specimen.

Table B1.2 Coefficients D_{nm} for eq.(B1.3.5)

m	n=0	1	2	3	4
0	2.673	-8.604	20.621	-14.635	0.477
1	-3.557	24.973	-53.398	50.707	-11.837
2	1.230	-8.411	16.957	-12.157	-0.940
3	-0.157	0.954	-1.284	-0.393	1.655

B1.3.3 Double cantilever beam (DCB)

The stress intensity factor for the DCB-specimen (Fig. B1.4) loaded at the crack mouth $x=0$ is

$$\frac{K_1}{P/B} = \sqrt{\frac{12}{d}}\left(\frac{a}{d}+0.68\right)+\sqrt{\frac{2}{\pi a}}\,\exp\!\left(-\sqrt{12\frac{a}{d}}\right) \qquad (B1.3.7)$$

An approximation for the biaxiality ratio is within $0.1<d/a<0.6$

$$\frac{1}{\beta} \cong 0.681\frac{d}{a}+0.0685 \qquad (B1.3.8)$$

The weight function is

$$h = \sqrt{\frac{12}{d}}\left(\frac{a}{d}+0.68\right)-\sqrt{\frac{12}{d}}\frac{x}{d}+\sqrt{\frac{2}{\pi(a-x)}}\,\exp\!\left(-\sqrt{12\frac{a-x}{d}}\right) \qquad (B1.3.9)$$

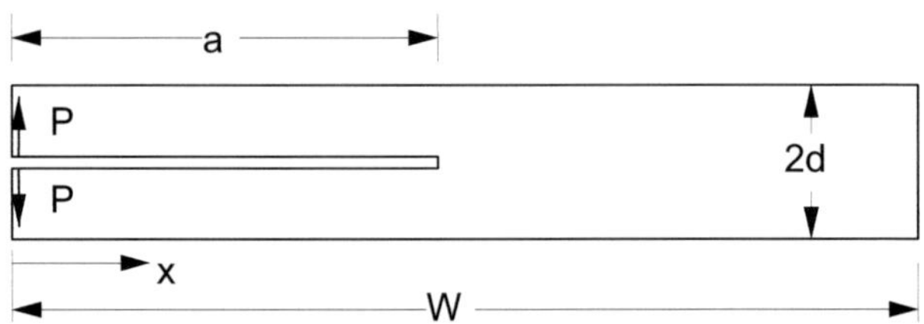

Fig. B1.4 Double Cantilever Beam specimen loaded by a line load P.

B1.3.4 Double cleavage drilled compression (DCDC) specimen

The stress intensity factor solution for the DCDC specimen (Fig. B1.5) was given by He et al. [B1.8] as

$$\frac{|p|\sqrt{\pi R}}{K_I} = \frac{1}{F} = \frac{H}{R}+\left[0.235\frac{H}{R}-0.259\right]\frac{a}{R} \qquad (B1.3.10)$$

20

The biaxiality ratio can roughly be expressed for $2 \leq a/R \leq 7$ and $H/R = 4$ by [B1.5]

$$\beta \approx -3\frac{a}{R} \qquad (B1.3.11a)$$

or for the extended region of $2 \leq a/R \leq 16$ by

$$\beta = -2.8\sqrt{\frac{a}{R}} - 0.678\left(\frac{a}{R}\right)^{3/2} \qquad (B1.3.11b)$$

The weight function is given as

$$h = \sqrt{\frac{2}{\pi a}}\left(\frac{1}{\sqrt{1-x/a}} + D_1\sqrt{1-x/a} + D_2(1-x/a)^{3/2}\right) \qquad (B1.3.12)$$

with the coefficients

$$D_1 = -3.97 + 1.46604\left(\frac{a}{R}\right) + 4.533\exp\left[-0.291\frac{a}{R}\right] \qquad (B1.3.13)$$

$$D_2 = -0.246 + 0.222\left(\frac{a}{R}\right) + 0.5288\exp\left[-2.08\frac{a}{R}\right] \qquad (B1.3.14)$$

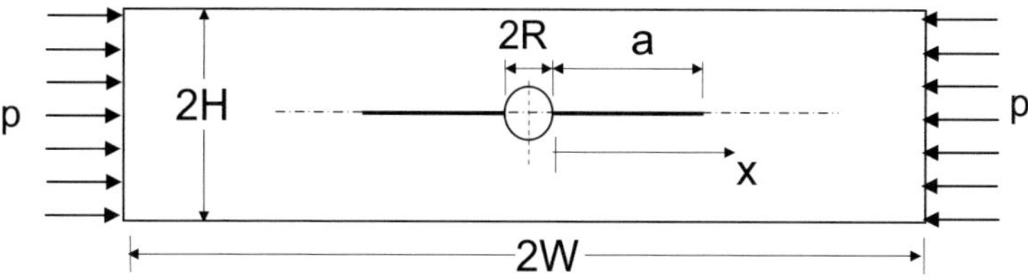

Fig. B1.5 Geometric data of a DCDC-specimen.

The relations mentioned before are valid for the standard geometry with $W/H=10$. For specimens with reduced lengths FE-computations had to be performed [B1.9]. The deviations from the standard solution K_0 obtained for the large "standard length" of $W_0=10H$ and $R=H/4$ are shown in Fig. B1.6. For a tolerated deviation of 1% at a crack length of $a/R<10$, the specimen length can be reduced to $W=0.45\,W_0$. A further reduction of the specimen length yields at least for the long crack of $a=10R$ rather strong deviations and should therefore be avoided. With other words the uncracked ligament of the bar $L=W-a-R$ should satisfy the condition of $L \geq 7/4\,H$.

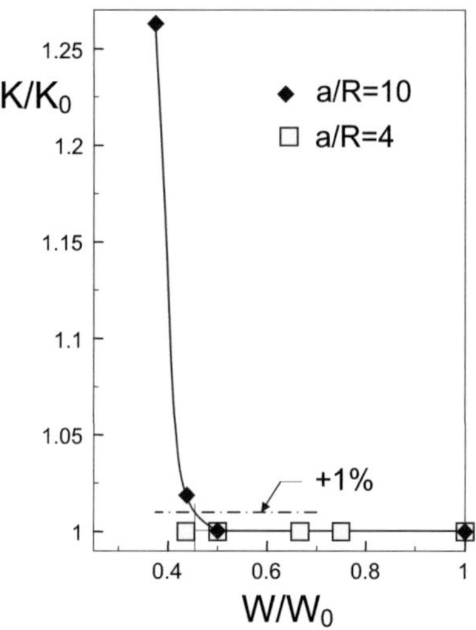

Fig. B1.6 Influence of the specimen length on the stress intensity factor (K_0=stress intensity factor for "standard geometry" W_0=10 H, H=4R).

B1.4 Crack opening displacement field

The relation of Rice [B1.10] allows to determine the weight function from a known crack opening displacement $\delta(x,a)$ under any arbitrarily chosen loading and the corresponding stress intensity factor $K_I(a)$ according to

$$h = \frac{E'}{K}\frac{\partial \delta}{\partial a}$$
(B1.4.1)

Here it has to be emphasized that the derivative in (B1.4.1) is only a partial variation of the displacements with crack length for a fixed loading, i.e. for a fixed stress distribution. By an inverse use, it also allows the crack opening displacement field to be determined for any load.

As outlined in [B1.4], integration of (B1.4.1) gives

$$\delta = \frac{1}{E'}\int_{a-r}^{a} h(r,a')da'\int_{0}^{a'} h(r',a')\sigma(r',a)dr'$$
(B1.4.2)

where r is the distance from the crack tip. As the consequence of the partial derivative, it should be emphasized that in the inner integral, the stress is $\sigma(r',a)$ for the <u>real</u> crack of length a and not the bridging stress distribution of a shorter crack of length a'.

B1.5 Compliance

The crack-length determination for crack resistance curves is often based on measurement of loading-point displacements δ_{LP} and load P where a record of $\delta_{LP}(P)$ sensitively indicates smallest crack growth.

The loading point displacements δ_{LP} and its increase $\Delta\delta_{LP}$ define the compliance C and the increment ΔC due to crack extension

$$C(a) = \frac{\delta_{LP}}{P}, \quad \Delta C(a) = \frac{\Delta\delta_{LP}}{P}. \tag{B1.5.1}$$

The compliance of a bending bar containing a crack of length a is given as

$$C = \frac{9}{2}\frac{L^2\pi}{BW^4E'}\int_0^a a'F^2(a')\,\mathrm{d}(a') + C_0 \tag{B1.5.2}$$

with the supporting span L in 3-point bending or $L=S_1-S_2$ in 4-point bending ($S_1=$ supporting roller span, $S_2=$ loading roller span) and the compliance C_0 of the bar without a crack. The function F is the geometric function for the stress intensity factor as defined by (B1.1.9).

B1.6 Cone crack

A cone crack in a semi-infinite body is illustrated in Fig. B1.7a. For a Hertzian contact pressure distribution

$$p(x) = p_0\sqrt{1-(x/a)^2} \tag{B1.6.1}$$

with maximum pressure p_0 in the centre, the mixed-mode stress intensity factors were determined with the FE method. K results from [B1.11] are plotted in Fig. B1.7b and B1.7c in terms of the geometric functions F_I and F_{II}

$$K_{I,II} = p_0 F_{I,II}\sqrt{c} \tag{B1.6.2}$$

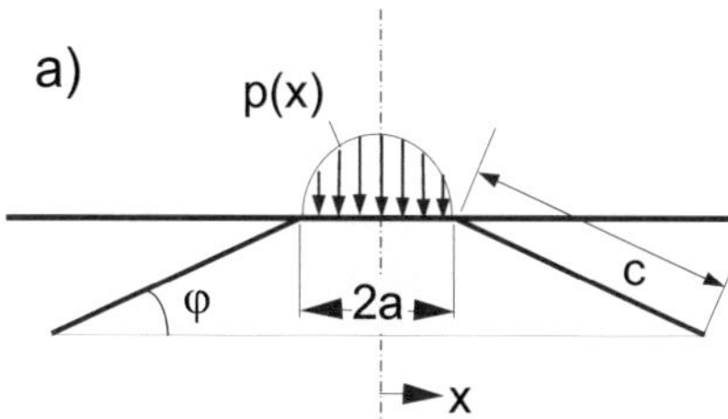

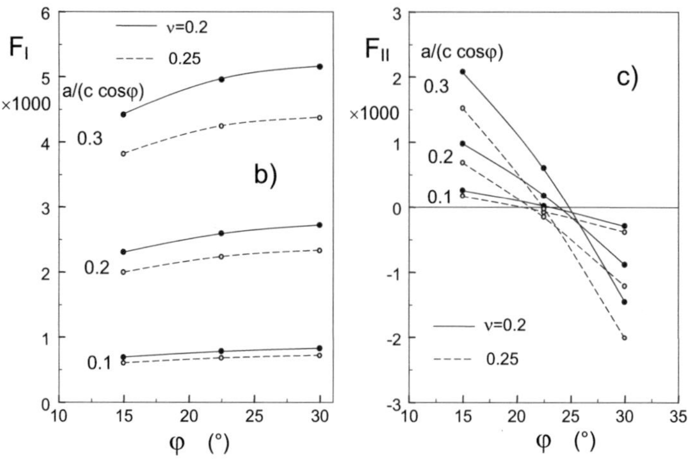

Fig. B1.7 a) cone crack, b), c) geometric functions for mixed-mode stress intensity factors [B1.11].

References B1

[B1.1] Williams, M.L., On the stress distribution at the base of a stationary crack, J. Appl. Mech. **24**(1957), 109-114.

[B1.2] Leevers, P.S., Radon, J.C., Inherent stress biaxiality in various fracture specimen geometries, Int. J. Fract. **19**(1982), 311-325.

[B1.3] Bückner, H., A novel principle for the computation of stress intensity factors, ZAMM **50** (1970), 529-546.

[B1.4] Fett, T., Munz, D., Stress intensity factors and weight functions, Computational Mechanics Publications, Southampton, 1997.

[B1.5] Fett T. Stress intensity factors, T-stresses, Weight function, IKM50, Universitätsverlag Karlsruhe, Karlsruhe; 2008.

[B1.6] Fett T. Stress intensity factors, T-stresses, Weight function (Supplement Volume), IKM55, KIT Scientific Publishing, Karlsruhe; 2009; (Open Access: http://digbib.ubka.uni-karlsruhe.de /volltexte/1000013835).

[B1.7] Newman, J.C., Stress analysis of compact specimens including the effects of pin loading, ASTM STP 560, 1974, 105.

[B1.8] He, M.Y., Turner, M.R., Evans, A.G., Analysis of the double cleavage drilled compression specimen for interface fracture energy measurements over a range of mode mixities, Acta metall. mater. **43**(1995), 3453-3458.

[B1.9] Rizzi, G., unpublished results, 2008.

[B1.10] Rice, J.R., Some remarks on elastic crack-tip stress fields, Int. J. Solids and Structures **8**(1972), 751-758.

[B1.11] Fett, T., Rizzi, G., Problems in fracture mechanics of indentation cracks, Report FZKA 6907, Forschungszentrum Karlsruhe, 2003, Karlsruhe.

B2 Equations for 2-dimensional cracks

B2.1 Weight functions

B2.1.1 Embedded circular crack

In the case of a stress distribution depending on the radial coordinate r exclusively, the stress intensity factor along the crack front of a circular crack in an infinite body (Fig. B2.1a) can be expressed in terms of a weight function by

$$K_{\mathrm{I}} = \int_0^a h(r)\sigma(r)dr \qquad (B2.1.1)$$

The related weight function is

$$h(r) = \frac{2r}{\sqrt{\pi a(a^2 - r^2)}} \qquad (B2.1.2)$$

This function is represented in Fig. B2.1b.

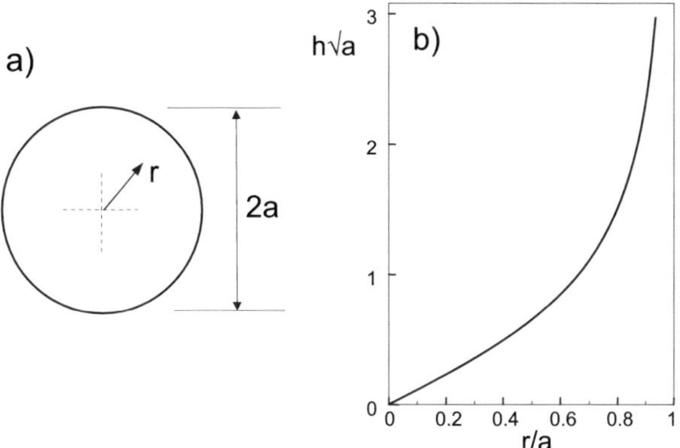

Fig. B2.1 a) Circular crack in an infinite body, b) weight function for the circular crack.

B2.1.2 Semi-circular surface crack

The weight function for a semi-circular surface crack (Fig. B2.2a) is

$$h(r,\varphi) \cong \frac{2r[1 + c(1 - r/a)]}{\sqrt{\pi a(a^2 - r^2)}} \qquad (B2.1.3)$$

where the coefficient c accounts for the influence of the free surface

$$c = \frac{0.04 + 0.104(1 - \sin \varphi)^2}{1 - \frac{\pi}{4}}$$ (B2.1.4)

The weight functions $h(r,\varphi)$ are plotted in Fig. B2.2b for $\varphi = 0$ and $\varphi = 90°$ as the thin lines. It can be seen that the influence of the angle φ is rather small. Consequently, it is recommended to neglect the angular influence by using in (B2.1.4) an average value of $c \cong 0.42$. This value results in the thick curve. The dash-dotted curve represents the solution for the embedded circular crack for comparison.

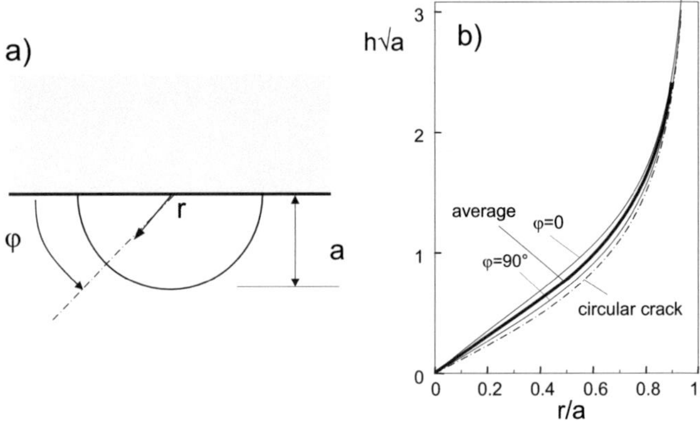

Fig. B2.2 a) Semi-circular surface crack, b) weight function for the semi-circular surface crack parallel to the surface ($\varphi = 0$) and normal to the surface ($\varphi = 90°$) compared with that for the circular crack in a semi-infinite body.

B2.2 Crack opening displacement field

Similar to eq.(B1.4.2), the crack opening displacement of a circular crack results by use of the weight function (B2.1.2) [B2.1]

$$\delta_{circ}(r) = \frac{4}{\pi E'} \int_r^a \left(\int_0^{a'} \frac{r'\sigma(r')}{\sqrt{a'^2 - r'^2}} \, dr' \right) \frac{da'}{\sqrt{a'^2 - r^2}}$$ (B2.2.1)

In an approximate extension, the crack opening displacement for the semi-circular crack can be computed from

$$\delta_{semi-circ}(r) \cong \frac{4}{\pi E'} \int_r^a \left(\int_0^{a'} \frac{r'(1 + c(1 - r'/a'))\sigma(r')}{\sqrt{a'^2 - r'^2}} \, dr' \right) \frac{(1 + c(1 - r/a'))da'}{\sqrt{a'^2 - r^2}}$$ (B2.2.2)

where the simplification $c= 0.42$ allows to handle the problem under conditions of rotational symmetry.

The circular and semi-circular cracks are often used as approximations for indentation cracks [B2.2]. As the applied load σ_{appl}, one can apply a residual stress field according to Hill [B2.3]

$$\sigma_{appl} = \sigma_{res} = \begin{cases} -p & \text{for } r < b \\ \tfrac{1}{2}p(b/r)^3 & \text{for } r > b \end{cases} \qquad (B2.2.3)$$

where r is the radial coordinate with origin in the crack centre as illustrated in Fig. B2.3.

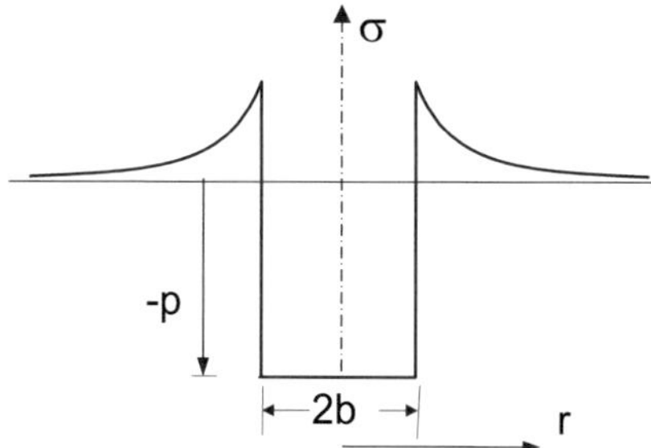

Fig. B2.3 Residual stresses by an expanding sphere [B2.3].

A semi-analytical solution for the COD field under the stress distribution given by eq.(B2.2.3) is

$$\frac{\delta_{appl}}{K_{appl}} = \frac{4\sqrt{a}}{0.382\,\pi E'}\left(\frac{a}{b}\right)^2\left[\frac{b}{2a}g_2(a,b,r) + (0.635 + 0.319\,b/a)g_1(a,b,r) - g_1(a,\lambda b,r)\right] \qquad (B2.2.4)$$

with

$$g_1(a,b,r) = \sqrt{1-(\tfrac{r}{a})^2}\,(1-\sqrt{1-(\tfrac{b}{a})^2}) + $$
$$+ \tfrac{r}{a}[\mathbf{E}((\tfrac{b}{r})^2) - E(\arcsin\tfrac{r}{a},(\tfrac{b}{r})^2) - (1-(\tfrac{b}{r})^2)(\mathbf{K}((\tfrac{b}{r})^2) - F(\arcsin\tfrac{r}{a},(\tfrac{b}{r})^2))] \qquad (B2.2.5)$$

$$g_2(a,b,r) = \tfrac{b}{r}[\mathbf{E}((\tfrac{b}{r})^2) - E(\arcsin(\tfrac{r}{a},(\tfrac{b}{r})^2)] \qquad (B2.2.6)$$

where $\mathbf{E}$ and $\mathbf{K}$ are the complete and $E(\ )$ and $F(\ )$ the incomplete elliptical integrals. The elastic modulus E' may be represented by the plane stress Yong's modulus E, since plane stress conditions prevail at the free surface. The geometric data are explained in Fig. B2.4a.

The parameter λ in (B2.2.4) reads

$$\lambda \cong 0.9828(a/b)^{0.00565} \tag{B2.2.7}$$

Equation (B2.2.4) was originally derived for the special case of an embedded circular crack. This relation can also be applied in very good approximation for a semi-circular surface crack, since in the ratio $\delta_\mathrm{appl}/K_\mathrm{appl}$ the individual influences of the different weight function on δ_appl and K_appl act in the same direction and are, therefore, nearly cancelled out.

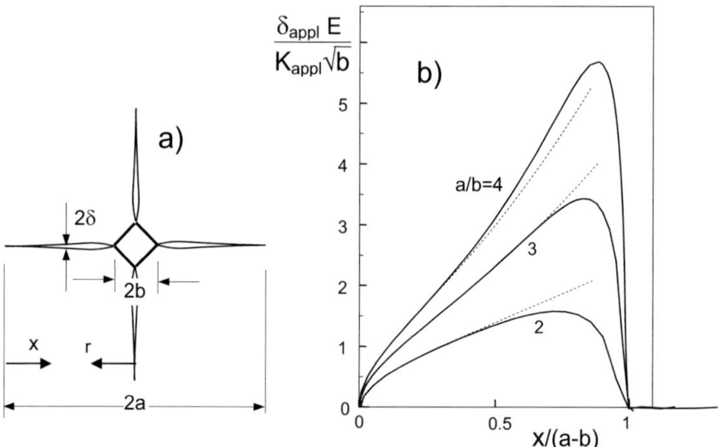

Fig. B2.4 a) Vickers indentation cracks (definition of parameters), b) crack opening displacements of Vickers indentation cracks: comparison of the analytical solution (solid curves) with the approximation eq.(B2.2.8) (dashed curves).

The displacement solution is shown in Fig. B2.4b by the solid curves. In [B2.4] a simplified series expansion for the displacements was proposed for $a/b > 1.4$. Considering the leading terms exclusively, it holds

$$\frac{\delta_{appl}}{K_{appl}} \cong \frac{\sqrt{b}}{E'}\left(\sqrt{\frac{8}{\pi}\frac{x}{b}} + A_1\left(\frac{x}{b}\right)^{3/2} + A_2\left(\frac{x}{b}\right)^{5/2}\right) \tag{B2.2.8}$$

with the coefficients A_1 and A_2 approximated by

$$A_1 \cong 11.7\exp[-2.063(a/b-1)^{0.28}] - \frac{0.898}{a/b-1}, \tag{B2.2.9}$$

$$A_2 \cong 44.5\exp[-3.712(a/b-1)^{0.28}]-\frac{1}{(a/b-1)^{3/2}} \qquad (B2.2.10)$$

A representation of the displacement approximation (B2.2.8) is given in Fig. B2.4b by the dashed curves.

B2.3 Weight function and near-tip COD evaluation

The weight function occurring in the displacement equations is available for many cracks. In the special case of the circular embedded crack h is given by eq.(B2.1.2) and an estimation for the semi-circular surface crack by eq. (B2.1.3).

Sometimes the question arises how accurately a weight function must be known for the determination of K_{I0} from near-tip crack opening displacements.

A series expansion of (B2.1.3) with respect to a-r yields

$$h_{semi-circ} = \sqrt{\frac{2}{\pi(a-r)}} + \frac{4c-3}{a\sqrt{8\pi}}\sqrt{a-r} + O((a-r)^{3/2}) \qquad (B2.3.1)$$

i.e. $\qquad h_{semi-circ} \cong h_{asympt}\,[1+(c-\tfrac{3}{4})(1-r/a)], \quad h_{asympt} \cong \sqrt{\frac{2}{\pi(a-r)}} \qquad (B2.3.2)$

This expansion also holds for the circular crack by setting $c=0$. It should be mentioned that the asymptotic term h_{asympt} is present in _any_ weight function for _any_ crack problem.

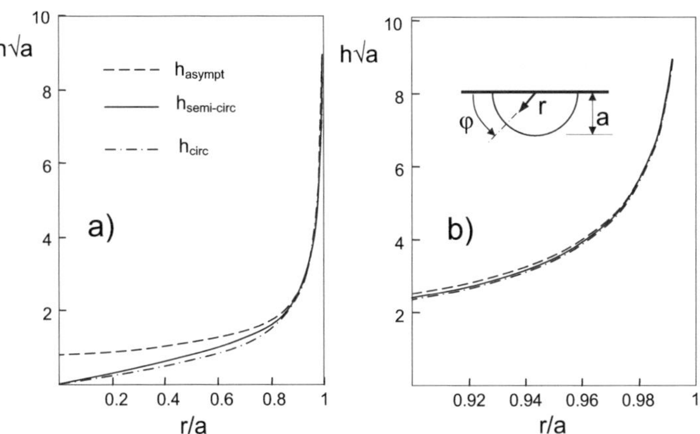

Fig. B2.5 a) Weight function for the semi-circular surface crack along the surface ($\varphi=0$) compared with the circular crack in a semi-infinite body and the asymptotic term of (B2.3.2), b) near-tip behaviour.

Figure B2.5 shows the weight functions according to eqs.(B2.1.2) and (B2.1.3) together with the asymptotic term of (B2.3.2). Whereas for $r/a<0.8$ clear differences between the "full" weight functions and the asymptotic term can be stated (Fig. B2.5a), a good agreement can be seen for $r/a>0.9$ (Fig. B2.5b).

For $a-r<0.05a$ the prevailing asymptotic weight function term with $1/(a-r)^{1/2}$ deviates less than 4% from the full solution. Consequently, the near-tip weight function is sufficiently represented by the singular term of (B2.3.2). Therefore, the near-tip crack opening displacements can simply be computed by application of relation (B1.4.2) for the straight-through crack (replacing the r-coordinate in (B1.4.2) by the x-coordinate used in this section).

References B2

B2.1 Sneddon, I.N., The distribution of stress in the neighbourhood of a crack in an elastic solid, Proc. of the Royal Soc., London, Ser. A **187**(1946), 229.

B2.2 Fett, T., Fünfschilling, S., Hoffmann, M. J., Oberacker, R., Different R-curves for 2- and 3-dimensional cracks, Int. J. Fract. **153**(2008), 153-159.

B2.3 Hill, R., Mathematical Theory of Plasticity, Oxford University Press, 1950, Oxford.

B2.4 Schneider, G.A., Fett, T., Computation of the stress intensity factor and COD for submicron sized indentation cracks, J. Ceram. Soc. of Japan, **114**(2006), 1044-1048.

B3 The problem of finite notch radii

B3.1 Slender notches as "starter cracks"

In Sections B1 and B2, the fracture mechanics relations were compiled which are necessary for the evaluation of R-curves. All these equations imply the pre-existence of a sharp crack. In most experimental investigations, however, slender starter notches are used. In earlier investigations starter notches were introduced as saw cuts, leading to notch root radii of about $R = 30 - 100$ μm. In recent years, rather thin notches could be produced with razor blades as proposed by Nishida et al. [B3.1]. This notching procedure yields notch roots with a radius of about $4 - 10$ μm

There are strong differences in the shape of the R-curve for different ceramics. Whereas coarse-grained alumina showed a moderate initial increase of K_R extending over crack growth in the order of several mm before reaching a plateau [B3.2], Si_3N_4-ceramics (MgO-La_2O_3- and MgO-Lu_2O_3-doped) exhibit a very steep initial increase occurring within only a few micrometers [B3.3, B3.4], comparable to the notch radius. Consequently, the evaluation of the initial parts of an R-curve under the assumption of a pre-existing crack is not possible at least for the steep R-curves.

Therefore, the fracture mechanics treatment of such notch problems may be explained briefly. A specimen containing a slender edge notch of depth a_0 with the notch root radius R is considered (Fig. B3.1a) with a small crack of length ℓ emanating directly at the notch root.

B3.2 Fracture mechanics relations for slender notches

B3.2.1 Stress intensity factor

In the first crack extension phase, the crack length ℓ may be comparable to R. The stress intensity factor K is then given by [B3.5]

$$K \cong K * \tanh[A \sqrt{\ell / R}] \tag{B3.2.1}$$

(A=2.243). The quantity $K*$ is the stress intensity factor of a crack of total length $a = a_0+\ell$ (see Fig. B3.1b). The related stress intensity factor (the so-called "long-crack approach") is given by

$$K* = \sigma\sqrt{\pi(a_0 + \ell)}\, F(a/W), \tag{B3.2.2}$$

where F is the geometric function for an edge crack of depth a in a specimen of width W.

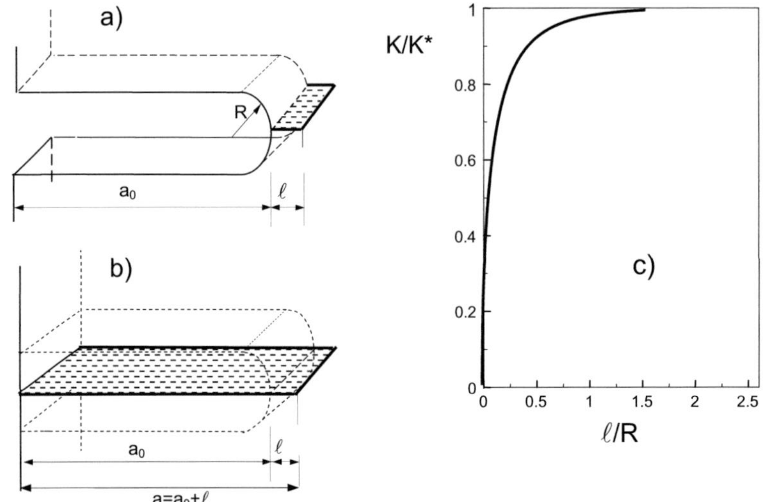

Fig. B3.1 a) Crack of length ℓ ahead of a slender notch with notch root radius R, b) same crack/notch configuration replaced by an auxiliary crack of total length $a=a_0+\ell$, c) stress intensity factor K normalized on the formally computed value K^*.

Equation (B3.2.2) is shown in Fig. B3.1c by the curve. From this plot it is clearly visible that the true stress intensity factor K is significantly lower than the formally computed K^* for the very first crack extension. On the other hand, it can be concluded that notch effects are negligible if $\ell > 1.5R$.

B3.2.2 T-stress

An approximate description for the T-stress is given by [B3.6]

$$\frac{T}{\sigma^*} \cong \frac{T_0}{\sigma^*} + \left(\frac{T^*}{\sigma^*} - \frac{T_0}{\sigma^*}\right) \tanh^{4/3}\left[\left(\frac{5\ell}{R}\right)^{3/4}\right] \tag{B3.2.3}$$

where σ^* is the characteristic stress (in our case the outer fibre bending stress). T^* is the T-stress term for the "long-crack solution", i.e. the T-stress for a crack of total length $a=\ell+a_0$ according to [B3.5]

$$\frac{T^*}{\sigma^*} = \frac{-0.526 + 2.481\,\alpha - 3.553\,\alpha^2 + 2.6384\,\alpha^3 - 0.9276\,\alpha^4}{(1-\alpha)^2} \tag{B3.2.4}$$

with $\alpha = a/W$ and

$$\frac{T_0}{\sigma^*} = -1.052 F(a_0/W)\sqrt{\frac{a_0}{R}} \qquad (B3.2.5)$$

B3.2.3 Weight function

For computations of bridging stress intensity factors for ceramics with crack-surface interactions it is necessary to know the weight function for the crack ahead of the notch root. For this purpose a simple interpolation relation was given in [B3.5] as

$$h_{notch} = \lambda h^{(2)} + (1-\lambda)h^{(1)} \quad , \quad \lambda = \left(\frac{R}{R+\ell}\right)^{7/2} \qquad (B3.2.6)$$

where
$$h^{(1)} = h_{edge}\left(\frac{\xi + a_0}{a}, \frac{a}{W}\right), \; \ell = a - a_0 \qquad (B3.2.7)$$

is the "long-crack weight function" for a crack of total length $a = a_0 + \ell$ and

$$h^{(2)} = h_{edge}\left(\frac{\xi}{\ell}, \frac{\ell}{W - a_0}\right) \qquad (B3.2.8)$$

is the weight function of an edge crack in a component of reduced width $W - a_0$.

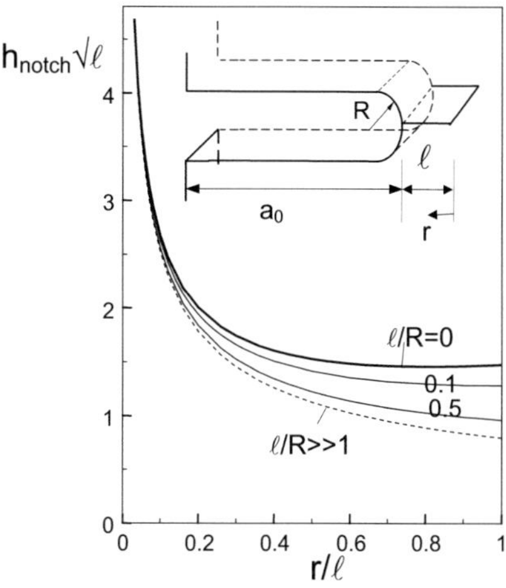

Fig. B3.2 Weight functions for a crack ahead of a slender notch.

The weight function (B3.2.6) reads for the special case of a small crack extension compared to the initial crack length, $\ell \ll a_0$

$$h_{notch} = \sqrt{\frac{2}{\pi \ell}} \left[\frac{1}{\sqrt{(\ell - \xi)/\ell}} + 0.568\,\lambda \left(\sqrt{\frac{\ell - \xi}{\ell}} + \tfrac{1}{2}\left(\frac{\ell - \xi}{\ell}\right)^{3/2} \right) \right] \qquad \text{(B3.2.9)}$$

with
$$\lambda = \left(\frac{R}{R+\ell}\right)^{7/2}, \quad \ell = a - a_0 \qquad \text{(B3.2.10)}$$

The result is illustrated in Fig. B3.2 for several ratios of ℓ/R.

B3.2.4 Compliance

In contrast to eq.(B1.5.2), the compliance of a crack in front of a slender notch is given by [B3.7]

$$\Delta C = \frac{9}{2} \frac{L^2 \pi}{B W^4 E'} \int_{a_0}^{a_0+\ell} (a_0 + \ell')\, F^2 (a_0 + \ell')\, \tanh^2[A\sqrt{\ell'/R}]\, \mathrm{d}(\ell') \qquad \text{(B3.2.11)}$$

with the supporting span L in 3-point bending or $L=S_1-S_2$ in 4-point bending ($S_1=$ supporting and $S_2=$ loading roller span). The compliance increment ΔC normalised on the formally computed value ΔC^* from (B1.5.2) is plotted in Fig. B3.3. From eq.(B3.2.11) it can be concluded that for crack extensions in the order of a few notch root radii ($\ell>4R$), the formally computed compliance will represent the notch/crack-configuration with sufficient accuracy.

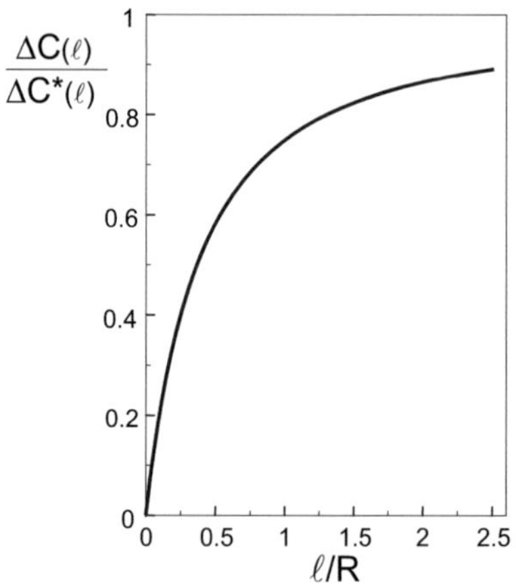

Fig. B3.3 True compliance increase ΔC normalised on the „long-crack solution" ΔC^* for the same crack increment.

References B3

B3.1 Nishida, T., Pezzotti, G., Mangialardi, T., Paolini, A.E., Fracture mechanics evaluation of ceramics by stable crack propagation in bend bar specimens, Fract. Mech. Ceram. **11**(1996) (Eds. R.C. Bradt, D.P.H. Hasselman, D. Munz, M. Sakai, V.Y. Shevchenko), 107–114.

B3.2 Steinbrech, R., Schmenkel,O., Crack resistance curves for surface cracks in alumina, Comm. J. Am. Ceram. Soc. **71**(1988), C271–C273.

B3.3 Satet, R. L., Hoffmann, M. J., Influence of the rare-earth element on the mechanical properties of RE-Mg-bearing silicon nitride, J. Am. Ceram. Soc. **88** (2005), 9, 2485–2490.

B3.4 Fett, T., Fünfschilling, S., Hoffmann, M.J., Oberacker, R., Jelitto, H., Schneider, G.A., R-curve determination for the initial stage of crack extension in Si_3N_4, J. Am. Ceram. Soc. **91**(2008), 3638-42.

B3.5 Fett, T., Munz, D., Stress Intensity Factors and Weight Functions, Computational Mechanics Publications, (1997) Southampton, UK.

B3.6 Fett T. Stress intensity factors, T-stresses, Weight function, IKM50, Universitätsverlag Karlsruhe, Karlsruhe; 2008.

B3.7 Fett, T., Notch effects in determination of fracture toughness and compliance, Int. J. Fract. **72**(1995), R27-R30.

C1 Investigated materials

C1.1 Silicon nitrides produced at IKM

Measurements at IKM were carried out on the following silicon nitrides. The materials were:

- A hot-isostatically-pressed silicon nitride with 5wt% Y_2O_3 and 2wt% MgO as sintering aids (denoted as MgY),

- hot-isostatically-pressed silicon nitride with 8.5 wt% Lu_2O_3 and 1.93 wt% MgO as sintering aids (denoted as MgLu),

- hot-isostatically-pressed silicon nitride with 7.1 wt% La_2O_3 and 1.96 wt% MgO as sintering aids (denoted as MgLa),

- hot-isostatically-pressed silicon nitride with 5wt% Y_2O_3 and 1.5 wt% SiO_2 as sintering aids (denoted as SiY).

Fünfschilling [C1.1] produced the materials investigated. They were consolidated in a two-step sintering process. The powder mixtures of silicon nitride (SN-E10, UBE) and additives were prepared by attrition milling in isopropanol and afterwards dried and sieved. Greenbodies (45mm × 64 mm × 6mm) were uniaxial pressed and subsequently cold-isostatically pressed. The samples were sintered in a hot-isostatic press. In the first step, the samples were sintered to achieve closed porosity and afterwards the HIP pressure of 10 or 20MPa was applied with a low N_2 pressure of 1MPa. Table C1.1 compiles the sintering conditions. For more details, see [C1.1].

Material	Sintering temperature, dwelling time, (°C, min)	HIP temperature, dwelling time, gas pressure (°C, min, MPa)	Average Grain Length (μm)	Peak value of aspect ratio
MgLa	1780/14	1780/30/20	≈1	9
MgLu	1750/15	1750/30/10	≈1	7
MgY	1750/60	1800/30/10	≈1	9
SiY	1920/25	1920/30/10	≈1	4

Table C1.1: Sintering conditions and microstructure.

The microstructure of the silicon nitrides is shown in Fig. C1.1. Characteristic data for the elongated β-crystals are entered in Table C1.1. These crystals often act as bridging events and are, therefore, responsible for the R-curve behaviour of Si_3N_4-ceramics.

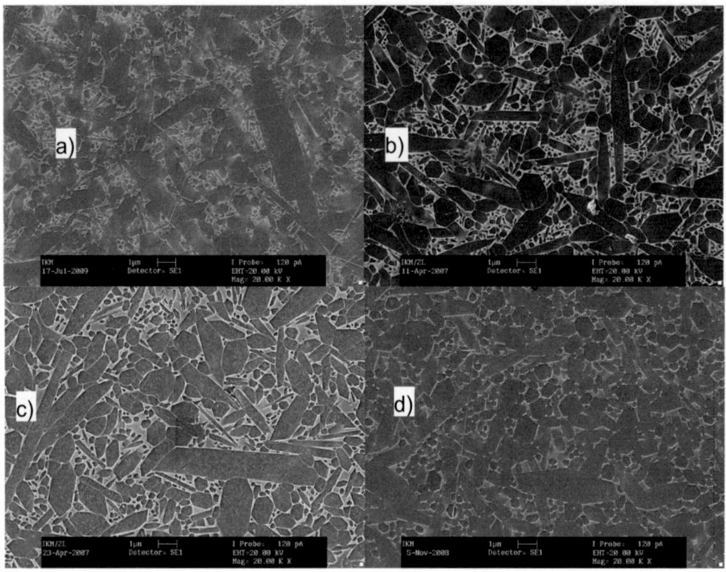

Fig. C1.1 Microstructure of the Si_3N_4-ceramics produced by Fünfschilling [C1.1], a) MgLa, b) MgLu, c) MgY, d) SiY.

Fünfschilling [C1.1] performed a statistical evaluation of the grain length as a function of the grain aspect ratio applying the procedure by Mücklich et al. [C1.2, C1.3]. Two examples of contour plots are given in Fig. C1.2 for MgY and SiY. Whereas the maximum frequency of the aspect ratio is about 9 for MgY, the SiY only shows rather short crystals with an aspect ratio of 3.5-4.

Some mechanics properties are given in Table C1.2.

Material	HV10 (GPa)	HK10 (GPa)	K_{Ic} (MPa√m)	K_{I0} (MPa√m)
MgLa	14.79±0.20	12.70±0.15	7.04±0.19	2.2±0.3
MgLu	15.31±0.15	13.50±0.06	6.48±0.14	2.2±0.3
MgY	15.06±0.26	13.63±0.09	6.77±0.28	2.2±0.3
SiY	14.88±0.20	12.78±0.08	3.59±0.78	2.2±0.3

Table C1.2: Properties of the Si_3N_4-ceramics (K_{I0} from Section D, K_{Ic} from SEVNB-tests).

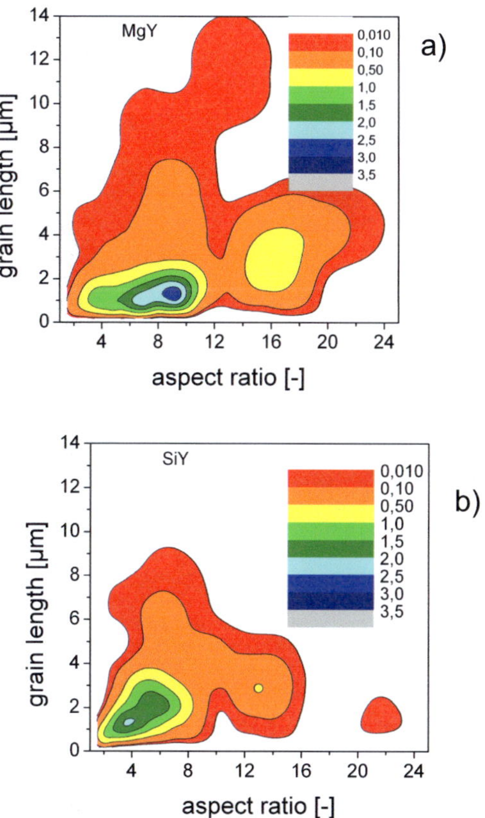

Fig. C1.2 Contour plots for the grain length as a function of the aspect ratio for the materials a) MgY and b) SiY (contour colours in%), taken from Fünfschilling [C1.1].

C1.2 Commercial silicon nitrides

Measurements similar to those on (Mg-Y)-containing Si_3N_4 were also carried out for several commercial silicon nitrides:

- a gas-pressured sintered silicon nitride FSN10 (FCT-Keramik, Rauenstein, Germany).

- SL200BG, a silicon nitride with 3wt%Y_2O_3 and 3wt% Al_2O_3 content (Ceram-Tec, Plochingen, Germany),

- the MgO-containing ceramic NC132 (Norton, Thousand Oaks, USA), and

- a gas-pressured silicon nitride containing 3% MgO (Ekasin-S, ESK, Kempten, Germany).

As a representative result, the microstructure for SL200BG is shown in Fig. C1.3.

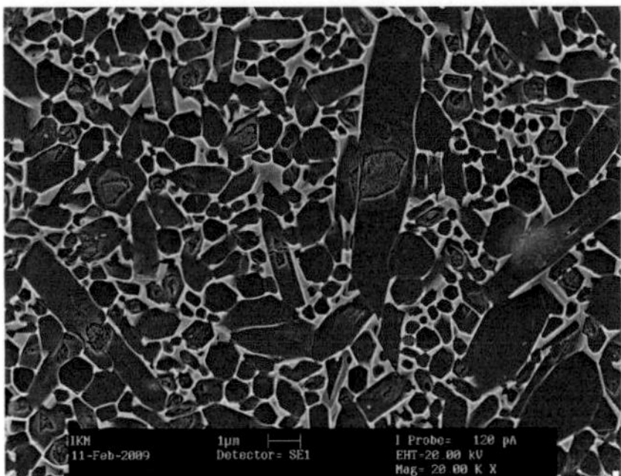

Fig. C1.3 Microstructure of SL200BG after [C1.1].

Material	HV10 (GPa)	HK10 (GPa)	K_{Ic} (MPa√m)	K_{I0} (MPa√m)
SL200BG	14.86±0.3	13.3±0.12	5.65±0.25	2.2±0.3
Ekasin-S	14	13.5 (HK1)	7	

Table C1.3: Properties of the commercial ceramics (from literature and data sheets of manufactures, data with standard deviations, by [C1.1]).

C1.3 SiAlON ceramics produced at IKM

SiAlON ceramics can be derived from both α- and β-silicon nitride modifications substituting silicon by aluminium and nitrogen by oxygen. The possibility to stabilize two crystal structures offers the opportunity to design materials ranging from pure α-, mixed α/β-, up to pure β- SiAlON. In general, α- SiAlONs are in the form of equiaxed grains with high hardness and good wear resistance but low fracture toughness, whereas β- SiAlONs have elongated grains with high fracture toughness but relatively low hardness. To combine the advantages of both SiAlONs, α/β-sialon composites have been developed over the past years by different mechanisms including the choice of starting powders, sintering parameters and the amount of additives. Sintering additives are necessary because SiAlONs are densified by a liquid phase sintering process like silicon nitride materials [C1.4].

In the work by Riva [C1.5], Nd_2O_3 was used as sintering additive in the starting powder mixtures for the fabrication of a mixed α/β- SiAlON [C1.6]. The manufacturing conditions of the SiAlONs were very similar to those for the silicon nitrides of Section C1.1.

The starting powders Si_3N_4, AlN, Al_2O_3, and Nd_2O_3 were attrition-milled for 4 h with silicon nitride milling balls in isopropanol. The slurry was subsequently dried in a rotator evaporator and finally sieved. Sintering of isostatically pressed plates was performed in nitrogen atmosphere in a BN-crucible using a hotisostatic press with a graphite resistance heater at a sintering temperature of 1830°C. During this heat-up phase, the pressure was kept below 1 MPa. When having reached the maximum temperature the pressure was held at 1 MPa for 15 min and then increased to 10 MPa for another 45 min (for more details see [C1.5]).

The compositions investigated were calculated according to the general SiAlON formula [C1.7]

$$M_x Si_{12-(m+n)} Al_{m+n} O_n N_{16-n}$$

with the coefficients $m = 0.5$ and 1.0 respectively, while n was kept constant at the value of 1.0. These compositions lay near the Si_3N_4 corner of the SiAlON plane as can be seen in Fig. C1.4.

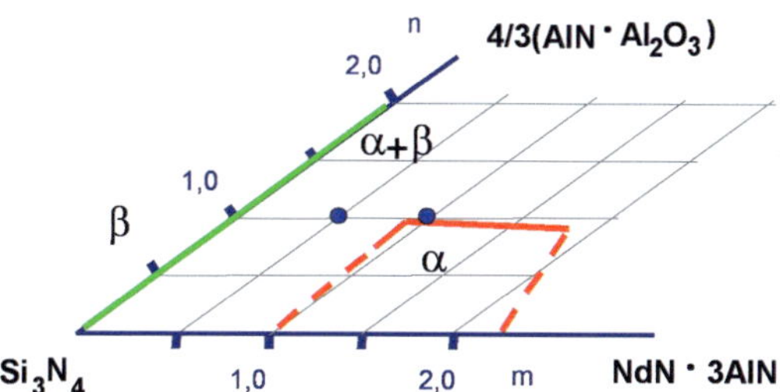

Fig. C1.4 Investigated compositions in the System Nd-Si-Al-O-N [C1.6]

Material	m	n	excess Nd_2O_3 [%]	α/β-ratio
Nd0510E1	0.5	1.0	10	55/45
Nd0510E6	0.5	1.0	60	55/45
Nd1010E1	0.5	1.0	10	90/10

Table C1.4 Nomenclature and composition of the investigated SiAlONs.

The so obtained plates exhibit a relative density of 99.4% – 99.8% of the theoretical density. The SEM micrographs in Fig. C1.5 show a balanced microstructure with a bright phase with an equiaxed grain morphology that corresponds to α-SiAlON and dark grains with elongated morphology corresponding to β-SiAlON. The small bright spots are rich in Neodymium and represent the small fraction of amorphous grain boundary phase located at triple junctions. No other crystalline phase could be deed by XRD after sintering. X-ray diffraction analysis of the samples showed an α/β-ratio of 55/45 for the compositions $m = 0.5$, $n = 1.0$ and 90/10 for the sample $m = n = 1.0$.

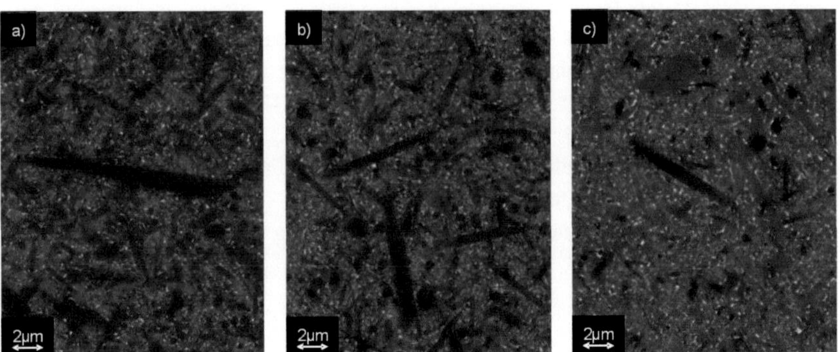

Fig. C1.5 SEM-micrographs of α/β-sialon, a) m=0.5, n=1.0 with 10%, b) 60% Nd_2O_3 (b) and c) α-rich SiAlON $m_{sialon}=n_{sialon}=1.0$.

In Table C1.5, mechanical properties of the investigated materials are summarized.

Material	HV10 (GPa)	E (GPa)	K_{Ic} (MPa√m)	K_{I0} (MPa√m)
Nd0510E1	19.62	319	5.54 ± 0.12	1.6
Nd0510E6	19.12	316	5.93 ± 0.04	1.5
Nd1010E1	20.85	321	5.22 ± 0.03	1.4

TableC1.5 Mechanical properties of the investigated SiAlONs (by Riva [C1.5]).

References C1

C1.1 Fünfschilling, S., Influence of micro-structure on the R-curve behaviour of silicon nitride ceramics (in German), Thesis, IKM 53, KIT Scientific Publishing, Karlsruhe, 2009; (Open Access: http://creativecommons.org.lizenses/by-nc-nd/3.0/de/).

C1.2 Mücklich, F., Ohser, J., Blank, S., Katrakova, D., Petzow, G., Stereological analysis of grain size and grain shape applied to silicon nitride ceramics, Z. Metallkunde **90**(1993), 557-561.

C1.3 Mücklich, F., Hartmann, S., Hoffmann, M.J., Schneider, G.A., Ohser, J., Petzow, G., Quantitative description of Si_3N_4 microstructures, Key Engng. Mat. **89-94**(1994), 465-470.

C1.4 Shen, ZJ, Ekström, T, Nygren, M, J. Am. Ceram. Soc. **79**(1996) 721.

C1.5 Riva, M., Entwicklung und Charakterisierung von SiAlON- Keramiken und SiAlON-SiC-Verbunden für den Einsatz in tribologisch hochbeanspruchten Gleitsystemen, IKM 54, KIT Scientific Publishing, Karlsruhe, 2011.

C1.6 Holzer, S, Geßwein, H, Hoffmann, MJ, Key Engng. Mat. **237**(2003), 43.

C1.7 Ekström, T., Nygren, M., J. Am. Ceram. Soc. **75**(1992), 259.

C2 R-curves for silicon nitride ceramics

C2.1 Measurements on notched bending bars

C2.1.1 R-curves from compliance evaluation

For the evaluation of $K_R(\Delta a)$, the actual crack length, a, has to be determined. Two possibilities are widely used for this purpose:

a) Direct optical observation of the crack on a side surface of the test specimen using a microscope;

b) Measurement of the specimen compliance and computation of the actual crack length via linear-elastic relations valid for traction-free cracks.

Both of these methods exhibit serious problems in practice:

- The *optical method* is applicable in all cases where the crack propagation test starts from a pre-existing crack [C2.1]. In most investigations, however, slender starter notches are used. At the beginning of a test, such notches exhibit no observable cracks at the side surfaces although the compliance clearly indicates crack extension [C2.2].

- The *compliance method* is correct at the beginning of a test ($\Delta a=0$). However, due to the generation of bridging interactions during crack propagation, the actual specimen stiffness is higher than that for a specimen containing a traction-free crack of same length. Consequently, the crack length from an evaluation of compliance measurements using the traction-free elastic compliance must result in an apparent crack length lower than the physical length, i.e. the elastic compliance method underestimates the real crack length.

In [C2.3], these two effects were discussed in detail based on experimental results on silicon nitride ceramics exhibiting steeply rising R-curves.

For the computation of the actual crack length ℓ (Fig. B3.1) from ΔC, eq.(B3.2.11) has to be solved with respect to the upper integration limit. During notching, a small initial crack of length, ℓ_0, may be generated by machining [C2.4, C2.5, C2.6], which during increasing load extends by an amount of $\Delta\ell$. It then holds for the total length:

$$\ell = \ell_0 + \Delta\ell . \tag{C2.1.1}$$

The effective initial crack size, ℓ_0, can be estimated by comparing the intrinsic crack tip toughness, K_{I0}, with the stress intensity factor computed for the load σ_0 at first crack extension (indicated by P_0 in Fig. C2.1). From eqs.(B3.2.1) and (B3.2.2) it results:

$$K_{I0} = \sigma_0 F(\tfrac{a_0+\ell_0}{W})\sqrt{\pi(a_0+\ell_0)}\,\tanh(A\sqrt{\ell_0/R}), \tag{C2.1.2}$$

If K_{I0} has been obtained independently from the near-tip crack opening displacement field as will be outlined in Section D2, consequently, the length ℓ_0 of the initial crack may be written as

$$\ell_0 \cong \frac{R}{A^2}\left(\tanh^{-1}\left[\frac{K_{I0}}{\sigma_0 F(a_0/W)\sqrt{\pi a_0}}\right]\right)^2 \tag{C2.1.3}$$

C2.1.2 Compliance evaluation including the effect of bridging tractions

In section B1.5, the compliance for free crack faces was considered. In order to include the influence of the bridging interactions on the compliance, a modified procedure has to be used [C2.3, C2.7].

By application of Betti's reciprocal work theorem, the displacements at the loading points caused by the bridging effect, $\delta_{LP,br}$, result as:

$$\tfrac{1}{2B} P_{appl}\delta_{LP,br} = \int_0^{\Delta a} \sigma_{br}(r)\delta_{appl}(r)\,dr , \tag{C2.1.4}$$

with the crack-tip distance r. The factor $1/(2B)$ appears since the total applied force, P_{appl}, in 4-point bending is split into <u>two</u> line loads over thickness B at the two loading points. The origin of the coordinate r is chosen here to be located at the crack tip. The "applied" crack-opening displacements, δ_{appl}, are defined as the displacements under the same load P_{appl} in the absence of the bridging stresses.

Application of eq.(B1.4.2) results in

$$\delta_{appl} = \frac{1}{E'}\int_{a-r}^{a} h(r,a')K_{appl}(a')\,da', \tag{C2.1.5}$$

where h is now the fracture mechanics weight function for a <u>crack ahead of the notch,</u> and K_{appl} the applied stress intensity factor.

In principle, the solution of eq.(C2.1.4) may be found iteratively. In the first approximation, the bridging effect on the compliance is ignored and the R-curve determined via the elastic compliance. Using this result in the form $K_R=K_{appl}$, the first approximation of the bridging law, $\sigma_{br}=f(\delta)$, can be determined by solving the simultaneous integral equations for the bridging stress intensity factor, K_{br}:

$$K_{I0} = K_{appl} + K_{br} = K_R + K_{br}, \qquad K_{br} < 0, \tag{C2.1.6}$$

$$K_{br}(\Delta a) = \int_0^{\Delta a} h(r,a)\,\sigma_{br}(\delta(r,a))\,dr \ , \tag{C2.1.7}$$

For the upper integration limit also $a=a_0+\Delta a$ can be used, because the bridging stresses disappear outside Δa, see (C5.2.3)).

The total displacements, δ:

$$\delta = \delta_{appl} + \delta_{br}, \tag{C2.1.8}$$

$$\delta_{br} = \frac{1}{E'} \int_{a-r}^{a} h(r,a')\,da' \int_0^{a'} h(r',a')\,\sigma_{br}(\delta(r',a))\,dr' \ , \tag{C2.1.9}$$

and the "applied displacements" according to eq.(C2.1.5):

$$\delta_{appl} = \frac{1}{E'} \int_{a-r}^{a} h(r,a')\,da' \int_0^{a'} h(r',a')\,\sigma_{appl}(r')\,dr', \tag{C2.1.10}$$

where σ_{appl} is the applied stress in the uncracked specimen.

With the first approximations of δ_{appl} and σ_{br}, the loading point displacements due to the bridging stresses result from eq.(C2.1.4). The related new compliance yields a new crack length and, consequently, an improved R-curve. Evaluation of this improved solution in the same way converges to the correct crack lengths and, finally, the correct R-curve. However, the procedure delineated before needs much effort and calls for an initial approximation (see Section C5.2).

C2.1.3 Example for the evaluation

The full R-curve procedure of section C2.1.2 may be applied to data for the commercial silicon nitride SL200BG. This material shows crack-tip toughness (the starting point of the R-curve) of $K_{I0}\cong2.2$ MPa√m (Section D2 and [C2.8]).

Four-point bending tests were carried out on notched bars of dimensions $W=4$ mm, $B=3$ mm, notch depths of about 2.5 mm and notch root radii in the range of

6µm≤R≤10µm (for experimental details see [C2.9]). The crack lengths were determined from load-displacement measurements applying the compliance method and were additionally measured with a microscope. Figure C2.1a shows the load-displacement curve P = f(δ_{LP}) near the region of the onset of crack extension.

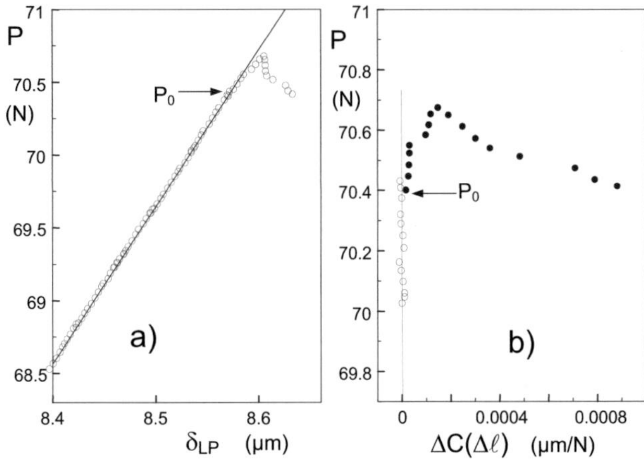

Fig. C2.1 Evaluation of experimental data for the silicon nitride SL200BG, a) individual data points compared with a straight-line fit of the $P(\delta)$-curve over the load range of 67 N < P < 69.5 N, b) load vs. change of compliance ΔC.

From a straight-line fit of the data in the load range of 67 N < P < 69.5 N, the linear "load vs. displacement" line in Fig. C2.1a was obtained. For any individual data point (δ_i,P_i), the difference $\Delta\delta_i$ with respect to the fit-line was determined by $\Delta\delta_i = \delta_i$-$\delta_{fit}$ and the change of the elastic compliance caused by the crack extension $\Delta\ell$=Δa computed as $\Delta C_i(\Delta\ell)$=$\Delta\delta_i/P_i$ and plotted in Fig. C2.1b. It should be mentioned that the crack length by microscopic inspection could not be identified before a crack length of about 25 µm was reached, whereas the compliance indicates crack lengths in the order of 1µm.

C2.1.4 Representation by an appropriate relation

By application of the procedure described in Sections C2.1.1 and C2.1.2, the R-curve data K_R were determined for all the materials represented in Section C1.1. The results are compiled in Fig.C2.2. These R-curves are initially very steep, as has been reported for Si_3N_4 ceramics with similar microstructures by Kruzic et al.[C2.10].

The data of Fig. C2.2 was best fitted to the expression:

$$K_R = K_{I0} + \sum_{i=1}^{N} A_i (1 - (1 + B_i \sqrt{\Delta a}) \exp[-B_i \sqrt{\Delta a}]) \qquad \text{(C2.1.11)}$$

Equation (C2.1.11) is shown in Fig C2.2b by the solid curves. Some of the coefficients are compiled in Table C2.1.

Material	A_1 (MPa√m)	B_1 (1/√m)	A_2 (MPa√m)	B_2 (1/√m)	K_{I0} (MPa√m)
MgLa	3.07	1742	1.63	581	2.2±0.3
MgLu	5.0	1542			2.2±0.3
MgY	4.65	1780	2.22	88.9	2.2±0.3
FSN10	4.12	1967	2.92	98.3	2.2±0.3

Table C2.1: Coefficients for the R-curves of Fig. C2.2 represented by eq.(C2.1.11); K_{I0} data from [C2.11].

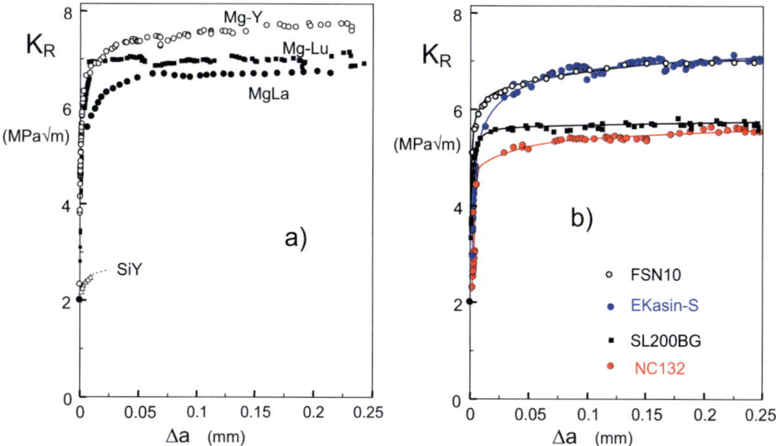

Fig. C2.2 R-curves for the materials of Sections C1.1 and C1.2.

C2.1.5 Differences in crack lengths between compliance and optical measurement

In the experiments mentioned before, the actual crack length was determined by evaluation of the increasing compliance. In addition, the optically visible crack length at the side surface was measured. Whereas for very short crack extensions no crack could be observed optically, the compliance clearly indicated crack extension. A typical result is shown in Fig. C2.3. It can be clearly seen that the optical crack-length measurements indicated by the circles yield about 20µm shorter crack increments, Δa, for the first part of crack extension compared with the lengths via compliance.

Possible reasons for the different crack lengths are discussed in [C2.3, C2.7]. The location at the notch front at which a crack will be initiated and propagate, is that of maximum $K(a_0)$ for the 3-dimensional crack problem.

It is well known in fracture mechanics that the stress intensity factor K and the energy release rate $G \propto K^2$ vary along a straight crack front. This fact is illustrated in Fig. C2.4a, where the local energy release rates G_{3D} are plotted normalised on the G-values obtained by 2D modelling assuming plane stress or plane strain conditions.

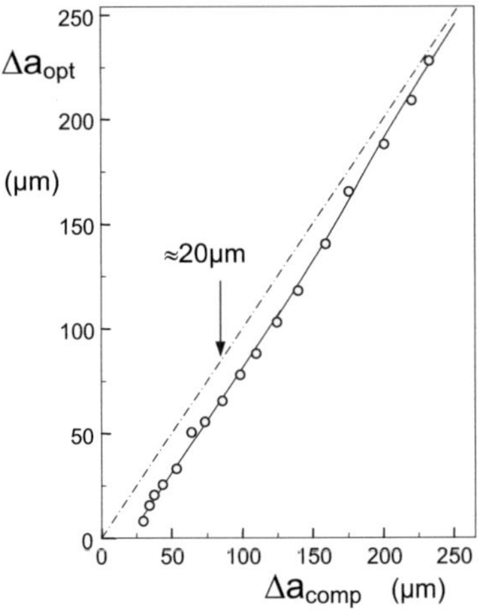

Fig. C2.3 Circles compare crack length measurement data collected via compliance and optical microscope while the dash-dotted line indicates perfect correspondence between the two measurements; the deviation is seen to be ~ 20 µm.

The squares show results of Dimitrov et al. [C2.12] obtained for a straight crack in a 3-point bending bar. The circles are results for a "double cleavage drilled compression" (DCDC) test specimen [C2.13]. In both cases, the energy release rates show a maximum in the specimen centre and significantly reduced values in the surface region.

Directly at the free side surface, $z/B \rightarrow \pm 1/2$, the description of the singular stress field by a stress intensity factor is no longer possible. In this case, the general relation represents the near-tip stresses

$$\sigma_y \propto r^{\lambda-1} \tag{C2.1.12}$$

with $\lambda \cong 0.54$ for a crack terminating angle of $\gamma=0$ (straight crack) [C2.14]. The singularity exponent λ depends on the crack terminating angle γ and on Poisson's ratio v. Equation (C2.1.12) yields a weak singularity for the stresses $\sigma \propto r^{-0.46}$ with the consequence of a disappearing energy release rate (for details see e.g. [C2.12]). In Fig. C2.4, this result is symbolised by the arrows (note that the finite G-values at the surface are a consequence of the finite FE-mesh). A finite energy release rate, necessary for stable crack growth, is ensured only if $\gamma \cong 10°$ where now $\lambda = \frac{1}{2}$ is fulfilled and, consequently, a stress intensity factor exists. With other words: A crack cannot grow stably at a free surface if $\gamma \neq 10°$.

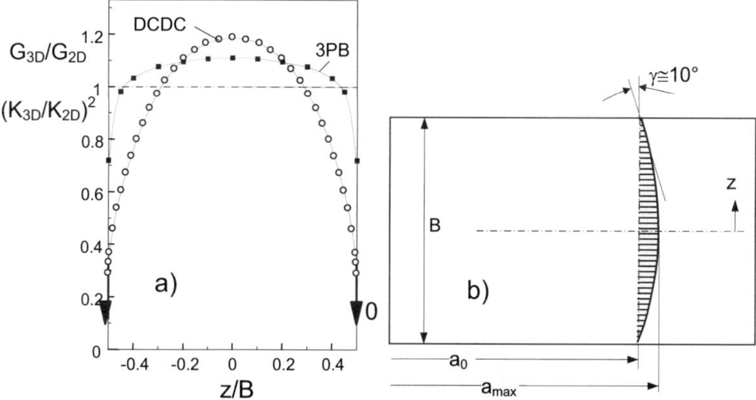

Fig. C2.4 a) Energy release rate distribution along the front of straight-through specimen cracks (squares: 3-point bending test [C2.12], circles: DCDC test specimen [C2.13]), both results obtained from FE modelling, b) first crack development in the centre region of the notch; when an angle of $\gamma \cong 10°$ is reached, crack extension is visible also at the side surface.

From this point of view, it can be concluded that first crack propagation must occur near the specimen center where the highest driving forces (G, K) are present (Fig. C2.4a). In a later crack extension phase the crack reaches the surface and can grow also in this region. Only now the crack becomes visible at the side surface and can be observed with the microscope.

C2.2 High-temperature R-curves from strength measurements

The actual state for fracture toughness measurements with starter notches is the use of the single edge V-notched bending (SEVNB) method [C2.15] (see also [C2.16, C2.17, C2.18]). With this procedure relatively small notch root radii of about 3-4 µm are reached. However, toughness measurement with V-notches still exhibits difficulties if the notch root radius is not small or very small compared with the mean grain size. Only, if the notch root radius is much less the mean grain size, the pre-existence of a "sharp" starter crack is always ensured.

A crack in a component starts to propagate when the externally applied stress intensity factor K_{appl} exceeds the so-called crack-tip toughness K_{I0}. For materials with a sufficiently steep R-curve, stable crack extension follows under increasing load. Failure of the component then occurs when the so-called tangent condition is fulfilled, i.e. when the slope of the $K_{appl}(a)$ and the R-curve $K_{IR}(\Delta a)$ are identical

$$K_{appl} = K_{IR} \; , \quad \frac{dK_{appl}}{da} = \frac{dK_{IR}}{da} \tag{C2.2.1}$$

The K_{IR} value at failure is called "fracture toughness" K_{Ic}.

From the failure condition eq.(C2.2.1), it follows that the K_{Ic} value (i.e. the applied stress intensity factor K_{appl} at failure) must be different for small and large cracks. For the assessment of small cracks as introduced in ceramic surfaces by grinding, it would be of advantage to know the initial part of the R-curve with crack extensions comparable to the crack size, i.e. for crack extensions in the order of about $\Delta a < 30\mu m$.

A macroscopic specimen for measurement of K_{Ic} was developed in [C2.19] containing a starter notch in the order of $a < 100\ \mu m$ depth with a notch root radius of $R < 100\ nm$ that was introduced by using a focussed ion beam (FIB) microscope. Due to limitations on the total sputtering time and the scanning width of the FIB-device, specimen thicknesses were limited to about 0.6 mm and maximum notch depths in the order of $a < 100\ \mu m$.

In [C2.19], a bending bar with a trapezoidal cross-section was chosen having a thickness at the tensile side of about 0.5mm, whereas the thickness at the compression surface was identical with the standard thickness of rectangular bending bars $\approx 3mm$. Since the trapezoidal cross section needs much effort in grinding, modified samples were developed.

C2.2.1 Initial R-curve from strength measurements on rectangular surface cracks

In order to generate a well-defined starter notch with a depth in the size of the natural crack population and a negligible notch-root radius, a rectangular surface crack of depth a and width $2c$ can be used as illustrated in Fig. C2.5a. Such a crack-like notch can be produced by using the FIB procedure.

For the rectangular crack, the stress intensity factors at points (A) and (B) under tensile loading are

$$K_{A,B,tension} = \sigma F_{A,B} \sqrt{\pi a} \tag{C2.2.2}$$

with the geometric function $F_{A,B}$ given in [C2.7]. Minimum stress intensity factors are present at the corners (C).

Based on tabulated data reported by Isida et al. [C2.20] (open circles in Fig. C2.5b), a fit-relation was derived in [C2.7] for the deepest point

$$F_A = \frac{F_0}{\left(1 + \frac{25}{8}\left(\frac{a}{c}\right)^{7/4}\right)^{0.276}}$$

(C2.2.3)

The value F_0 denotes the well-known solution for the edge crack, given by $F_0 = 1.1215$. This stress intensity factor solution is introduced in Fig. C2.5b as the solid line.

The solid circles in Fig. C2.5b are numerical data by Noguchi and Smith [C2.21] for the geometric function F_B. The square represents a result by Murakami [C2.22]. These F_B-values together with the trivial solution of $F_B=0$ for $a/c=0$ (edge crack) show that the stress intensity factor at point (A) is the maximum value only for roughly $a/c<3/4$. From this point, it is recommended to produce notches with a/c as small as possible.

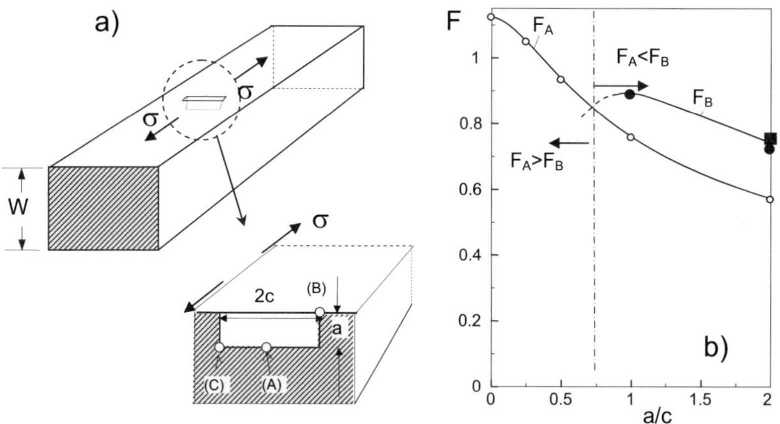

Fig. C2.5 a) FIB-introduced rectangular crack under tension, b) geometric functions at point (B) and point (A) according to [C2.7, C2.24]; open circles: geometric functions for the stress intensity factor at point (A) by Isida et al.[C2.20], solid circles: stress intensity factors at point (B) by Noguchi and Smith [C2.21], square: result by Murakami [C2.22].

C2.2.2 Application to high-temperature R-curve measurement on Si₃N₄

It is in general not possible to determine the fracture toughness K_{Ic} from only one single test because the crack length at failure is in principle unknown and cannot be identified after the test in all cases by crack surface inspection. By application of an indirect method it is possible to determine the initial, strength-relevant part of the R-curve, and, consequently, the K_{Ic} in dependence of crack length [C2.23], [C2.24].

Figure C2.6a illustrates the method as proposed in [C2.24], restricted to a series of for instance 3 specimens with notches of different depths tested in bending. The strength σ_c defines the related K_{appl} curve by using σ_c as the bending stress at failure. These K_{Ic} values represent the tangent points between K_{appl} and the R-curve indicated by the solid circles. Consequently, the inner envelope of the series of K_{appl} curves must describe the failure-relevant part of the R-curve. If the K_{I0}-value (open circle in Fig. C2.6a) is known too, the R-curve can be extended to $\Delta a \rightarrow 0$. For room temperature it is recommended to perform COD measurements for the K_{I0}-determination (see e.g. [C2.25, C2.26] and Section D2).

Since bending strength tests can easily be performed even at very high temperatures, the failure-relevant part of the R-curve can also be determined at elevated temperature. In [24], one test was made with a FIB-introduced surface notch (given by the $K_{appl}(a)$ curve (1)). Two additional tests were performed on differently deep notches generated via the razor blade technique resulting in the curves (2) and (3). The lowest curve segments were then joined to one polygonal curve. Since the K_{appl}-curves define the envelope, the R-curve must touch each of the single K_{appl} curves in one point and otherwise must lie below the envelope. If we smooth the polygonal curve at the crossing points, an improved approximation of $K_R(\Delta a)$ is obtained. For an appropriate smoothing of the polygonal curve, the room temperature R-curve (Fig. C2.6b) can be used as a template.

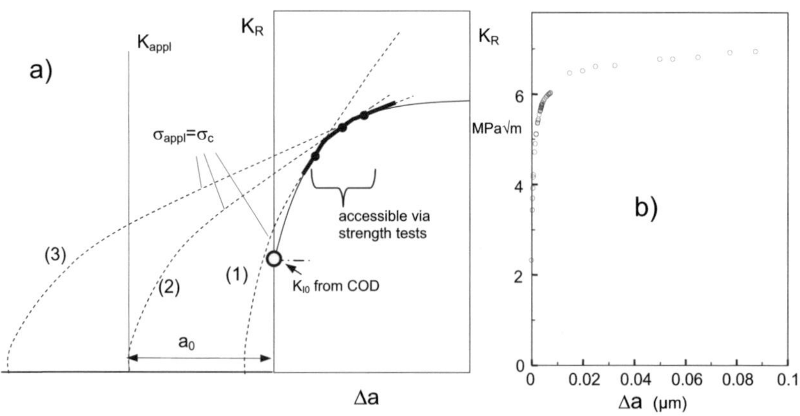

Fig. C2.6 a) Initial part of the R-curve from strength tests with FIB-introduced notches, construction of an upper limit for the R-curve (thick polygonal curve) by putting the lowest segments of the different K_{appl} curves together, b) initial room temperature R-curve from Section C2.1.

In [C2.24] the experiments on the silicon nitride FSN10 were carried out at elevated temperature. In order to illustrate the method, results obtained at 900°C were reported. This temperature is relevant for first crack extension under thermal shock and fatigue loading of rolls for wire hot rolling at multi-line rolling mills [C2.27].

The V-notched bars of $W = 4$ mm width representing the "long-crack behaviour" showed notch root radii in the range of $R = 6$-10 μm. The notch depths were chosen as $a_0 = 600$μm and 1900 μm. The "short-crack" region was realized by application of an edge-crack-like rectangular surface FIB-notch of dimensions a=5.3μm and $2c$=370μm showing notch-root radii of less than 0.1μm which are clearly smaller than the average grain size.

The FIB-generated samples we tested in 3-point bending in order to reduce the effective surface and volume responsible for failure at natural flaws and to ensure failure at the FIB-introduced notches. Nevertheless, two test specimens failed at natural flaws and had to be excluded from evaluation.

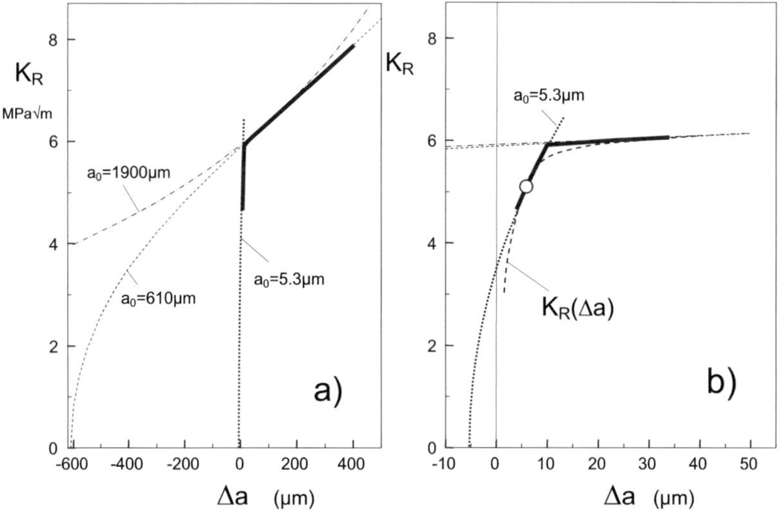

Fig. C2.7 a) Estimated high-temperature R-curve; thick lines: upper limit for the R-curve, thin curves: K_{appl}-curves for differently long starter notches, b) details of a) near Δa=0.

The final result is shown in Fig. C2.7. Figure C2.7a represents the K_{appl}-curves for the three tests as the thin lines. The envelope for K_R resulting from these curves is entered as thick polygonal curve. In Fig. C2.7b this is illustrated in more detail. The dashed curve, $K_R(\Delta a)$, smoothing the intersecting region near Δa=10µm is tentatively introduced by using the general shape obtained from the curve measurements at room temperature. Unfortunately, the high-temperature value of K_{I0} is still not available.

Whereas for the V-notched bending bars a fracture toughness of about 6-6.3 MPa√m was determined, the toughness for small strength-relevant cracks was clearly reduced to $K_{Ic}\cong$5.1 MPa√m as indicated in Fig. C2.7b by the circle representing the tangent point.

C2.3 Toughness value K_{I0} from COD-measurements on edge-cracks

The determination of K_{I0} from growing cracks needs much effort. In this context an evaluation proposed by Kounga et al. [C2.28] may be quoted. The sintered reaction-bonded silicon nitride (SRBSN) investigated in this study was a commercial material containing yttria and alumina. This SRBSN had needle-like beta silicon nitride grains, 0.5 µm to 3 µm wide by up to 10 µm long, bonded by a second phase. Standard rectangular bend specimens of size 3 mm x 4 mm x 45 mm were sliced and ground. A V-notch was cut on the edge of each specimen using a razor blade.

The samples were mounted in a device for bending tests and then loaded until a sharp crack was initiated. The testing device was placed under an optical microscope, in a manner that measurements could be carried out on the polished side of the specimen. After achievement of certain crack extensions, the specimens were removed from the testing device and introduced in a sputtering device. A very thin Au/Pd layer was deposited on the polished sample surface that facilitated further (SEM) measurements on the insulating material surface.

An external load was applied to the specimen thus enabling the crack to be opened up to the critical state of crack propagation as obtained in the optical microscope. To avoid significant sub-critical crack growth during the measurement of the crack profile, the externally applied load was reduced by 3%, 10%, and 20%, respectively. The entire crack profiles were recorded and, subsequently, the crack opening displacements measured. These data were then extrapolated into crack opening displacements related to the full load as applied during the prior stable crack extension phase.

Figures C2.8a, C2.8b, and C2.8c show a series of COD profiles. The curves represent an average of the individual data points. In Fig. C2.8d a comparison of the averaged dependencies is given. A fit of the results summarised in Fig. C2.8d shows that the total crack opening displacement δ_{total} in the presence of the applied stress intensity factor K_{appl} is related to the maximum crack opening displacement at maximum applied stress intensity factor $K_{appl,max}$ by

$$\frac{\delta}{\delta_{max}} = (1 - B) + B \frac{K_{appl}}{K_{appl,max}} \tag{C2.3.1}$$

with the coefficient B represented in Fig. C2.9a.

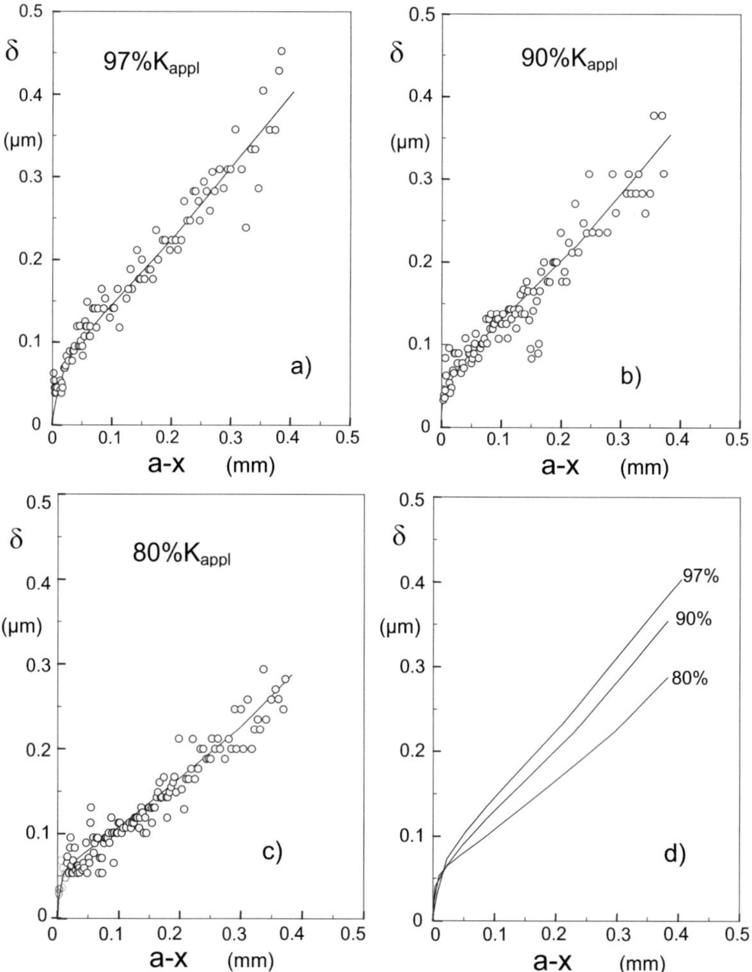

Fig. C2.8 Crack opening displacements for a grown crack after different load removals.

As an example of this procedure, the crack opening displacements for a specimen after crack propagation, measured at 90% load, are plotted in Fig. C2.9b as the squares. Application of eq.(C2.3.1) yields the COD at full load represented by the circles. These final results were the basis of the K_{I0} evaluation (see also Section D2). An evaluation of the data from [C2.28] resulted in $K_{I0} \cong 1.70 \pm 0.25$ MPa√m.

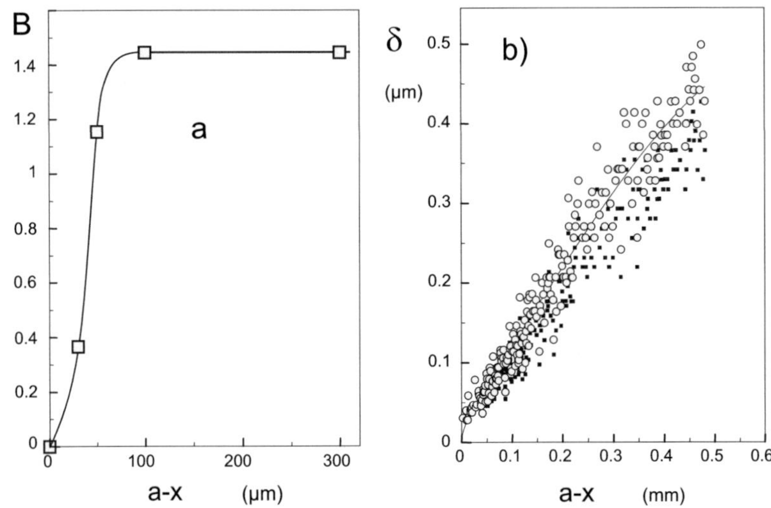

Fig. C2.9 a) Coefficient B for eq.(C2.3.1) obtained from Fig. C2.8, b) crack opening displacement for maximum load (circles) computed from measurements performed at 90% load (squares).

References C2

C2.1 Kruzic, J.J., Cannon, R.M., Ager III, J.W., Ritchie, R.O., Fatigue threshold R-curves for predicting reliability of ceramics under cyclic loading, Acta Mater., **53**(2005), 2595-2605.

C2.2 Fett, T., Influence of bridging stresses on specimen compliances, Engng. Fract. Mech., **53**(1996), 363-370

C2.3 Fünfschilling, S; Fett, T; Oberacker, R; Hoffmann, MJ; Oezcoban, H; Jelitto, H; Schneider, GA; Kruzic, JJ, R-curves from compliance and optical crack-length measurements, J. Am. Ceram. Soc., **93**(2010), 2814-21.

C2.4 Damani R., Gstrein R., Danzer R., "Critical Notch-Root Radius Effect in SENB-S Fracture Toughness Testing", J. European Ceramic Society **16**(1996), 695-702.

C2.5 Damani, R.J., Schuster, C., Danzer, R., Polished notch modification of SENB-S fracture toughness testing, J. European Ceramic Society, **17**(1997), 1685-1689.

C2.6 Fett, T., Munz, D., Influence of narrow starter notches on the initial crack growth resistance curve of ceramics, Archive of Applied Mechanics, (2006) DOI 10.1007/s00419-006-0055-3, pp. 1-13.

C2.7 Fett, T., Stress intensity factors, T-stresses, weight functions, -Supplement Volume-, Universitätsverlag Karlsruhe, Volume IKM55 (2009), (Open Access: http://digbib.ubka.uni-karlsruhe.de/volltexte/1000013835).

C2.8 Fünfschilling, S., Fett, T., Hoffmann, M.J., Oberacker, R., Jelitto, H., Schneider, G.A., Determination of the crack-tip toughness in silicon nitride ceramics, J. Mater. Sci. Letters, **44**(2009), 335-338.

C2.9 Fett, T., Fünfschilling, S., Hoffmann, M.J., Oberacker, R., Jelitto, H., Schneider, G.A., R-curve determination for the initial stage of crack extension in Si_3N_4, J. Am. Ceram. Soc., **91**(2008), 3638-42.

C2.10 Kruzic, J. J., Satet, R. L., Hoffmann, M. J., Cannon, R. M., Ritchie, R. O., The utility of *R*-curves for understanding fracture toughness-strength relations in bridging ceramics, J. Am. Ceram. Soc., **91**(2008), 1986-1994.

C2.11 Fünfschilling, S., Fett, T., Hoffmann, M.J., Oberacker, R, Schneider, G.A., Becher, P.F., Kruzic, J.J., Crack-tip toughness from Vickers crack-tip opening displacements for materials with strongly rising R-curves, J. Am. Cer. Soc., **94**(2011), 1884–1892.

C2.12 Dimitrov, A., Buchholz, F.-G., Schnack, E., "3D-corner effects in crack propagation." In H.A. Mang, F.G. Rammerstorfer, and J. Eberhardsteiner, eds, *On-line Proc. 5th World Congress in Comp. Mech. (WCCMV)*, Vienna, Austria, July 7–12(2002); http://wccm.tuwien.ac.at.

C2.13 Fett, T., Rizzi, G., Guin, J.P., López-Cepero, J.M., Wiederhorn, S.M., A fracture mechanics analysis of the DCDC test specimen, Engng. Fract. Mech., **76**(2009), 921-934.

C2.14 Benthem, J.P., "State of stress at the vertex of a quarter-infinite crack in a half-space," Int. J. Solids and Struct. **13**(1977), 479.

C2.15 Nishida, T., Pezzotti, G., Mangialardi, T., Paolini, A.E., Fracture mechanics evaluation of ceramics by stable crack propagation in bend bar specimens, Fract. Mech. Ceram., **11**, (1996), 107–114.

C2.16 Damani R., Gstrein R., Danzer R. ,"Critical Notch-Root Radius Effect in SENB-S Fracture Toughness Testing", J. European Ceramic Society, **16**(1996), 695-702.

C2.17 Damani, R.J., Schuster, C., Danzer, R., Polished notch modification of SENB-S fracture toughness testing, J. European Ceramic Society, **17**(1997), 1685-1689.

C2.18 Kübler, J., Fracture toughness using the SEVNB method: Preliminary results, Ceramic Engineering & Science Proceedings, **18**(1997), 155-162.

C2.19 Fett, T., Creek, D., Wagner, S., Rizzi, G., Volkert C., Fracture toughness test with a sharp notch introduced by focussed ion beam, Int. J. Fract. **153**(2008), 85-92.

C2.20 Isida, M. Yoshida, T., Noguchi, H., A rectangular crack in an infinite solid, a semi-infinite solid and a finite-thickness plate subjected to tension, Int. J. Fract., **52**(1991), 79-90.

C2.21 Noguchi, H., Smith, R., An analysis of a semi-infinite solid with three-dimensional cracks, Engng. Fract. Mech., **52**(1995), 1-14.

C2.22 Murakami, Y., Analysis of stress intensity factors of mode I, II, III of inclined surface crack of arbitrary shape, Engng. Fract. Mech., **22**(1985), 101-114.

C2.23 Pompe, W., Bahr, H.-A., Gille, G., Kreher, W., Schultrich, B., Weiss, H.J., Mechanical properties of brittle materials; Modern theories and experimental evidence, Current Topics in Materials Science, Vol. 12, North-Holland, 1985.

C2.24 Fünfschilling, S; Fett, T; Hoffmann, MJ; Oberacker, R; Oezcoban, H; Schneider, GA; Brenner, P; Gerthsen, D; Danzer, R., Estimation of the high-temperature R-curve for ceramics from strength measurements including specimens with focused ion beam notches, J. Am. Ceram. Soc., **93**(2010), 2411-14.

C2.25 Kounga Njiwa, A.B., Fett, T., Rödel, J., Quinn, G.D., Crack-tip toughness measurements on a sintered reaction-bonded Si3N4, J. Am. Ceram. Soc., **87**(2004), 1502-1508.

C2.26 Pezzotti, G., Muraki, N., Maeda, N., Satou, K., Nishida, T., In situ measurement of bridging stresses in toughened silicon nitride using Raman microprobe spectroscopy, J. Am. Ceram. Soc., **82**(1999), 1249-56

C2.27 Lengauer, M., Danzer, R.; Silicon nitride tools fort he hot rolling of high-alloyed steel and superalloy wires – Crack growth and lifetime prediction, J. Eur. Ceram. Soc., **28**(2008), 2289-2298.

C2.28 Kounga Njiwa, A.B., Fett, T., Rödel, J., Quinn, G.D., Crack-tip toughness measurements on a sintered reaction-bonded Si_3N_4, J. Am. Ceram. Soc., **87**(2004), 1502-1508.

C3 Bridging stresses in silicon nitride ceramics

C3.1 Computation of bridging stresses from the R-curve

It is the common opinion that in the special case of R-curves caused by grain bridging effects, the relation between the bridging stresses and crack opening displacement, $\sigma_{br} = f(\delta)$, is the intrinsic material property which is expected to be unaffected by test conditions as geometry of the test specimen or special type of loading (tension, bending, etc.).

Direct measurements of the loads transferred by the bridges were performed by Pezzotti et al. [C3.1, C3.2] and Kruzic et al.[C3.3] applying Raman spectroscopy. A further method to determine the bridging stress relation is the evaluation of crack opening displacement (COD) measurements [C3.4].

In this Chapter, the determination of bridging stresses from the R-curves is outlined in detail.

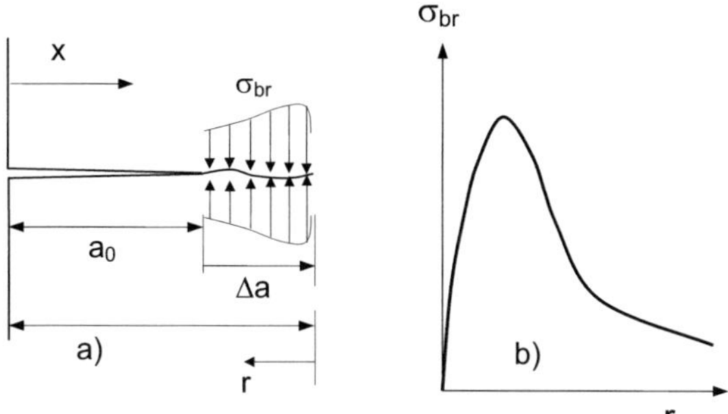

Fig. C3.1 a) A crack in a ceramic material exhibiting crack surface interactions by bridging stresses, b) bridging stress distribution versus crack-tip distance r.

A procedure that allows the bridging stresses to be determined from existing R-curve results was developed in [C3.5]. With this technique, originally applied to coarse-grained alumina, also R-curve measurements on silicon nitrides obtained by a high-resolution compliance method [C3.6] could be evaluated.

From the measured R-curves the bridging stress intensity factor, K_{br}, is known since

$$K_R = K_{I0} - K_{br} \, , \quad K_{br} < 0 \tag{C3.1.1}$$

(K_{I0}=crack-tip toughness) must hold during stable crack propagation. This fact allows the bridging stresses to be determined. The necessary procedures are extensively outlined in literature. For our purpose, we used the technique described in [C3.5].

First, we make use of the fracture mechanics insight that any stress intensity factor can be represented by a stress distribution acting in the wake of the crack. Using the weight function representation, the bridging stress intensity factor is given by

$$K_{br}(\Delta a) = \int_0^{\Delta a} h(r,a)\,\sigma_{br}(r)\,dr \qquad (C3.1.2)$$

with the fracture mechanics weight function h (for details see [C3.7, C3.8]), the distance r from the tip, the initial crack length a_0 free of bridging, and the crack extension $\Delta a = a - a_0$ (see Fig. C3.1).

For a given bridging stress distribution $\sigma_{br}(r)$, eq.(C3.1.2) gives the related bridging stress intensity factor. For the inverse problem of determination the bridging stresses from R-curve data, eq.(C3.1.2) has to be considered as an integral equation with the function $K_{br}(\Delta a)$ known and the bridging stress to be determined.

In most cases slender notches are used as "starter cracks". Consequently, the weight function for a small crack of length ℓ emanating from the notch root (for geometry see insert in Fig. C3.2) has to be used in eq.(C3.1.2). Such a solution was given in [C3.8] (for details see also Section B3.2.2). In the special case of a crack extension increment small compared to the initial crack length, $a - a_0 = \ell \ll a_0$, h can be written as

$$h_{notch} = \sqrt{\frac{2}{\pi\ell}}\left[\frac{1}{\sqrt{r/\ell}} + 0.568\,\lambda\left(\sqrt{\frac{r}{\ell}} + \tfrac{1}{2}\left(\frac{r}{\ell}\right)^{3/2}\right)\right] \qquad (C3.1.3a)$$

with
$$\lambda = \left(\frac{R}{R+\ell}\right)^{7/2}, \quad \ell = a - a_0 \qquad (C3.1.3b)$$

Figure C3.2 shows the weight function solution (C3.1.3a) together with the geometric data R, ℓ, and r. For very short crack extensions in the order of about $\ell \leq R$, the weight function significantly depends on ℓ/R.

Since the bridging stresses σ_{br} depend on opening displacements, δ, this quantity has to be known, in order to establish the bridging law. The total displacements can be computed as the sum of the "applied" and bridging displacements, δ_{appl} and δ_{br}, respectively

$$\delta = \frac{1}{E'}\int_{a-r}^{a} h(r,a')\,K_{appl}(a')\,da' + \frac{1}{E'}\int_{a-r}^{a} h(r,a')\left[\int_0^{a'-a_0} h(r',a')\,\sigma_{br}(r')\,dr'\right]da' \qquad (C3.1.4)$$

with the plane strain modulus $E'=E/(1-\nu^2)$, ν=Poisson's ratio.

The first integral term accounts for the "applied displacements" (the displacements under same load that would appear in the absence of bridging stresses) and the second for the bridging displacements.

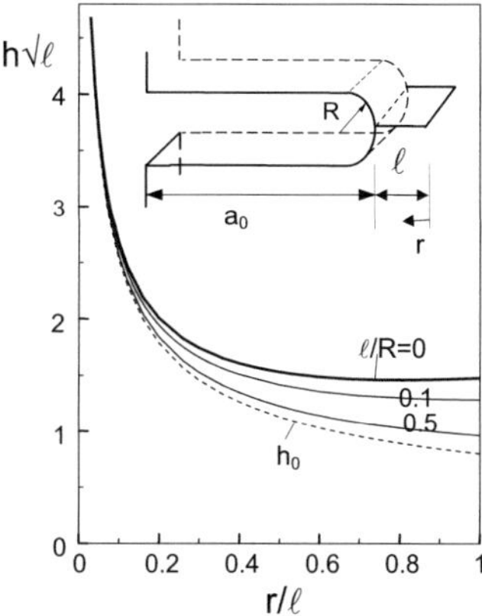

Fig. C3.2 Weight function for a crack ahead of a slender notch.

The individual R-curves of Section C2 could be fitted to the best by an expression of the type

$$K_R = K_{I0} + \sum_{(n)} A_n (1 - (1 + B_n \sqrt{\Delta a}) \exp[-B_n \sqrt{\Delta a}]) \tag{C3.1.5}$$

The general procedure of computing the bridging stresses may be described using the results of the (Mg-Y)-containing ceramic for which a crack-tip toughness of $K_{I0} \cong 2.2$ MPa√m was determined from COD-measurements [C3.9, C3.10]. The R-curve data obtained with a starter notch of $R = 6.2\mu m$ root radius are plotted in Fig. C3.3a by the symbols. These data were fitted according to (C3.1.5) resulting in the parameters A_1=4.65 MPa√m, B_1=1780/√m, A_2=2.22 MPa√m, and B_2= 88.9/√m (see Table C2.1). From this representation, a first-order solution for bridging stresses can be estimated according to [C3.11]

$$\sigma_{br}(r) \cong -\frac{1}{h} \frac{dK_R}{da}\bigg|_{\Delta a=r} \tag{C3.1.6}$$

as shown in Fig. C3.3b.

Approximation of the total near-tip displacements by the Irwin parabola for the near-tip behaviour ($r \to 0$)

$$\delta_{tip} = \sqrt{\frac{8r}{\pi}}\,\frac{K_{I0}}{E'} \qquad (C3.1.7)$$

allows a simple transformation from the r-dependence into the δ-dependence of stresses. By introducing (C3.1.5) and (C3.1.7) into (C3.1.6), it simply follows for the type of the bridging law

$$\sigma_{br}(\delta) = \sum_{(n)} C_n \delta \exp(-D_n \delta) \qquad (C3.1.8)$$

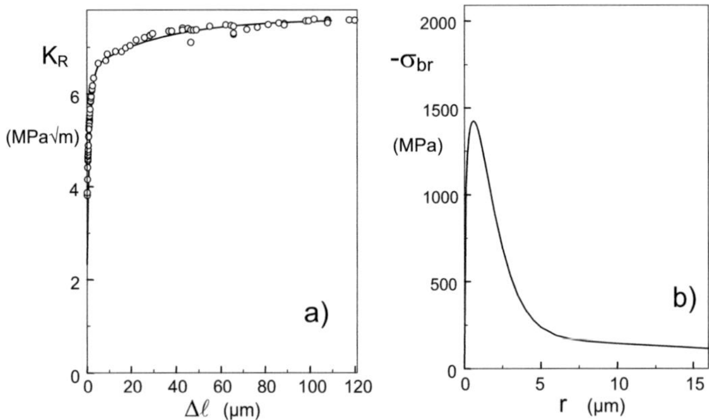

Fig. C3.3 a) R-curve for the Si_3N_4 with (Y_2O_3-MgO)-content; symbols: experimental results, curve: K_R fitted by eq.(C3.1.6), b) first-order bridging stress distribution near the crack tip according to eq.(C3.1.6).

Since the principal shape of the bridging stress relation is known from the first-order solution, numerical effort in determination of the "full" solution of bridging stresses can be drastically reduced. For this purpose the simple set-up for the unknown relation can be made

$$\sigma_{br} \approx \sum_{(n)} \sigma_n \frac{\delta}{\delta_n} \exp[-\delta / \delta_n] \qquad (C3.1.9)$$

This bridging stress relation has to be inserted into eqs.(C3.1.2) and (C3.1.4). Solution of this system of integral equations by systematic variation of the free parameters in (C3.1.9) under the condition that the resulting stress intensity factor K_R must coincide with the measured K_R-data provides the best set of unknown parameters σ_n and δ_n. This has been done for the materials listed in Section C1. Figures C3.4a and C3.4b show the near-tip distribution of the bridging stresses for the materials produced by Fünfschilling [C3.12] and the commercial materials, respectively.

Finally, combining $\sigma_{br}(r)$ and $\delta(r)$ yields the bridging law $\sigma_{br}(\delta)$ as plotted in Fig. C3.5a and C3.5b. The parameters for eq.(C3.1.9) are compiled in Table C3.1. In context with these data, it should be emphasized that the value σ_0 is not identical with the maximum bridging stress.

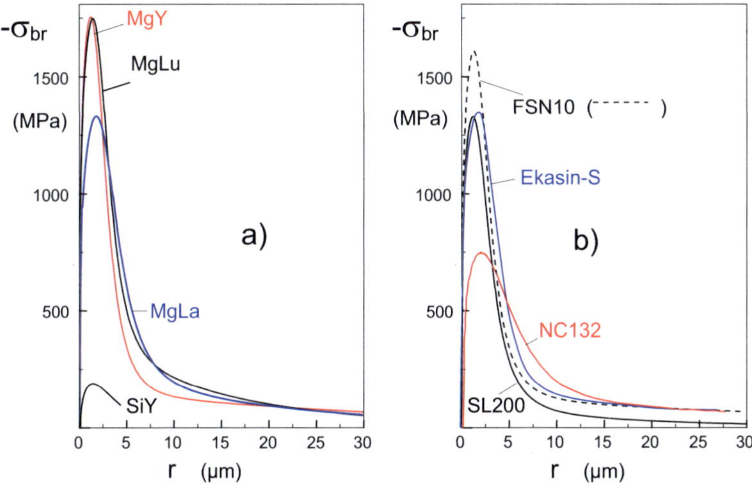

Fig. C3.4 a), b) Distribution of bridging stresses $\sigma_{br}(r)$.

Material	$-\sigma_0$ (MPa)	δ_0 (μm)	$-\sigma_1$ (MPa)	δ_1 (μm)	$-\sigma_2$ (MPa)	δ_2 (μm)
Mg-La	3375	0.0118	450	0.0476		
Mg-Lu	4370	0.0105	700	0.044		
Mg-Y	4600	0.0116	250	0.072	50	0.12
FSN10	4050	0.01	300	0.02	250	0.08
SL200	3500	0.01	200	0.037		
NC132	1900	0.0125	270	0.0555		
Ekasin-S	3650	0.0122	230	0.0625	60	0.2

Table C3.1: Coefficients for the bridging law equation (C3.1.9).

For a drastic reduction in numeric effort, the approximate procedure described in Section C5.2 was applied. The results are shown in Fig. C5.4. These solutions show the same ranking in maximum bridging stresses as the full solutions shown in Fig. C3.4.

There is obviously an influence of the aspect ratio of the β-crystals on the maximum stress values. Figure C3.5c shows the peak stresses of Fig. C3.5a versus the most frequent aspect ratios according to Fig. C1.2 and Table C1.1. The general trend is an increasing bridging stress with increasing aspect ratio.

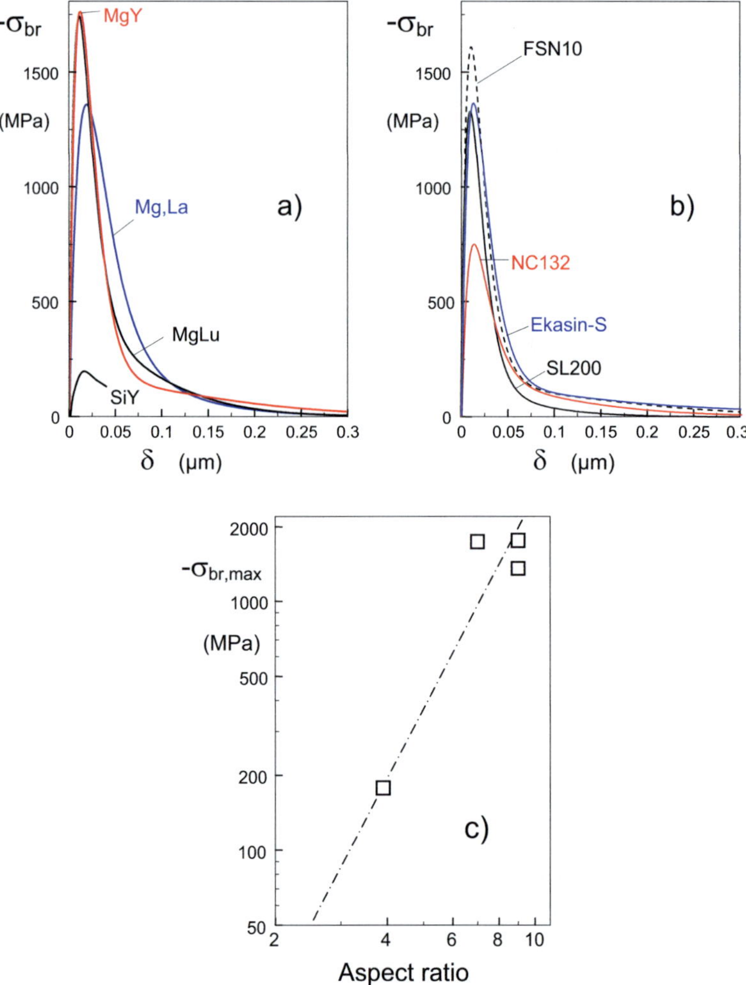

Fig. C3.5 a), b) Bridging laws $\sigma_{br}(\delta)$, c) influence of the most frequent aspect ratio on the peak value of the bridging stress.

C3.2 Interpretation of the bridging stresses

C3.2.1 Bulk material

The bridging stresses for small crack opening displacements are very high with maximum values in the order of $-\sigma_{br}$ = 1250-1750 MPa. An interpretation of this behaviour was tried in [C3.12,C3.13].

Local condition for crack propagation: The investigated SN-ceramics consist of 100% β-crystals with needle-like grains showing high aspect ratios. Such a needle with length axis oriented under an angle of Θ relative to the plane of a crack terminating at the elongated crystal is schematically shown in Fig. C3.6a.

The crack in Fig. C3.6a has in principle two possibilities to grow. In cases where the grain boundary strength is very high, fracture must occur through the β-crystal (Fig. C3.6b). If the strength of the grain boundary is lower, the crack must kink into the direction parallel to the boundary resulting in a debonding length ℓ as illustrated in Fig. C3.6c. The transition from propagation case Fig. C3.6b to that of Fig. C3.6c defines a critical angle Θ_{cr}. If $\Theta<\Theta_{cr}$ crack propagation along the grain boundary prevails. If $\Theta>\Theta_{cr}$ crack grows through the β-crystal.

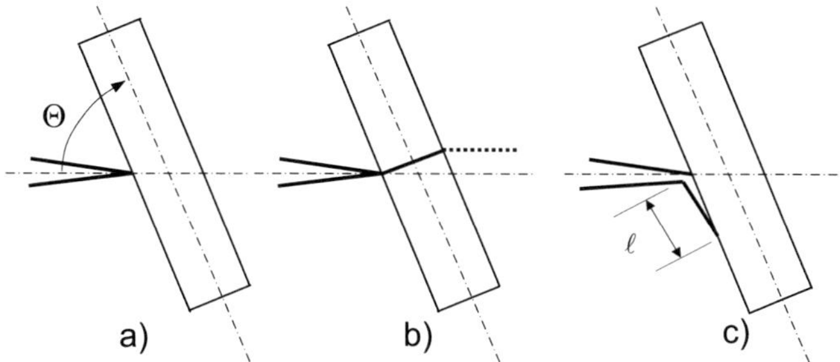

Fig. C3.6 a) Crack terminating an elongated β-crystal; b) crystal cracking in case of high grain-boundary strengths, c) crack kinking with following debonding for lower interface strength.

A method for an experimental determination of the critical angle between the grain-length-direction and the crack-plane direction was proposed by Becher et al.[C3.14]. Satet [C3.15] applied this method to La- and Lu-containing Si_3N_4-ceramics. The resulting debonding lengths are plotted in Fig. C3.7a versus the crack incidence angle. The case, in which any debonding disappears, $\ell\to0$, establishes the critical incidence angle Θ_{cr}. It results for Lu-containing SN $\Theta_{cr}=40°$ and for La-containing SN $\Theta_{cr}\approx60°$.

Percentage p of elongated β-needles which must fracture without any debonding, results to be

$$p = 1 - \sin\Theta_{cr} \tag{C3.2.1}$$

For Lu-containing SN with $\Theta_{cr}=40°$ it holds $p=0.36=36\%$. For this part of β-needles in the prospective crack plane (slightly more than are orientated in a statistical sense normal to the crack surface … namely 1/3), straight-through fracture must occur. For a La-SN, only 14% of elongated β-needles must fracture without any debonding.

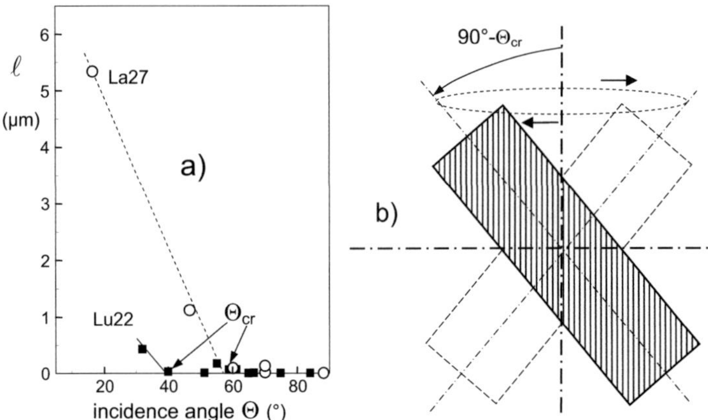

Fig. C3.7 a) Debonding length ℓ according to Fig. C3.6c versus crack terminating angle Θ for two Si₃N₄-ceramics with La and Lu-contents (measured by Satet [C3.15]), b) spherical angle with orientations for which no debonding occurs.

Global crack propagation:

The elastic constants of the SN-crystals on the one hand depend on the directions with respect to the c-axis. On the other hand, the elastic properties of the grain boundary layers may in general be different from those of the crystals. This must result in complex stress intensity factors, which are not vividly. In order to allow a description in terms of conventional stress intensity factors and toughness, let us ignore here the differences in the elastic parameters.

The portion $(1-p)$ of β-crystals which undergo crack extension along their grain boundary layers are now summarized and homogenized as a "low-toughness" matrix with toughness $K_{Ic,m}$ in which the portion p of crystals necessarily to be cracked during crack extension are embedded. These crystals with a higher toughness $K_{Ic,\beta}>K_{Ic,m}$ are obstacles for the free crack extension, which act as crack pinning events when the crack tries to pass. For reasons of simplicity, their orientation is chosen in the following illustrations to be normal to the crack plane.

In Fig. C3.8a a crack of initial length a_0 is shown coming from the left. The portion of β-crystals not able to debond is indicated by the dashed circles.

First crack extension takes place when the externally applied stress intensity factor K_{appl} equals the toughness of the matrix. Consequently, this value has to be identified as the so-called crack-tip toughness K_{I0}, the starting value of the R-curve.

When the crack front approaches the first pinning element, the crack must arrest there because $K_{Ic,\beta} > K_{appl}$. The crack front in the matrix material can propagate furthermore. This leads to a curvature of the initially straight crack front.

In the first-order analysis by Gao and Rice [C3.16] (for application to brittle materials see e.g. Roux et al. [C3.17]), the local stress intensity factor variation of a curved crack in an infinite body is given by

$$K_{curved}(z) = K_{straight}\left(1 + \frac{1}{\pi} \int\limits_{-\infty}^{\infty} \frac{\Delta a(z') - \Delta a(z)}{(z'-z)^2}\, dz'\right) \tag{C3.2.2}$$

with the coordinate direction of z defined in Fig. C3.9a and the applied stress intensity factor for a straight crack front $K_{straight}$. Note that the integrand in (C3.2.2) is a measure for the crack-front curvature.

From (C3.2.2) it follows that the stress intensity factor is reduced at locations where the crack could advance (convex crack front). At the pinning location where crack arrest is observed (concave crack front) K_{appl} is increased.

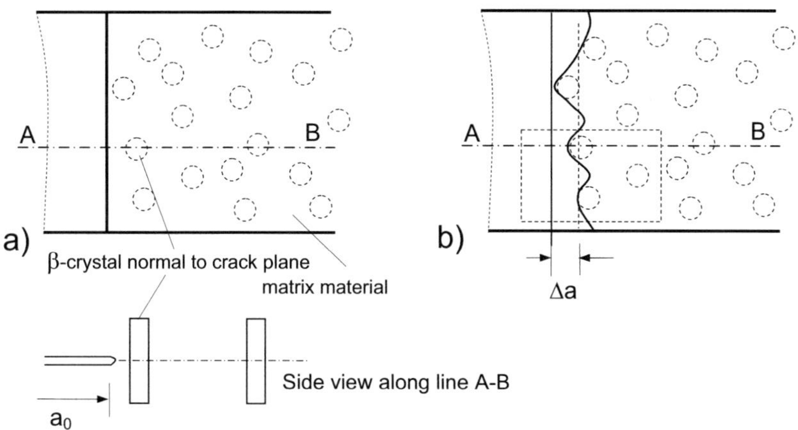

Fig. C3.8 Initial crack extension appearing in the "matrix" material, a) cross section in the uncracked material, b) crack extension with pinning at individual β-crystals.

After a certain amount of crack extension, the isolated β-crystals surrounded by a part of the crack front act as bridging events (Fig. C3.9b). There exists no free neck around the crystals because debonding did not occur before. Nevertheless, these regions carry tractions between the upper and the lower crack faces. The stress intensity factor at these locations increases not only with externally applied load, but also with increasing average crack opening caused by an increasing distance of the monotonously moving main crack front.

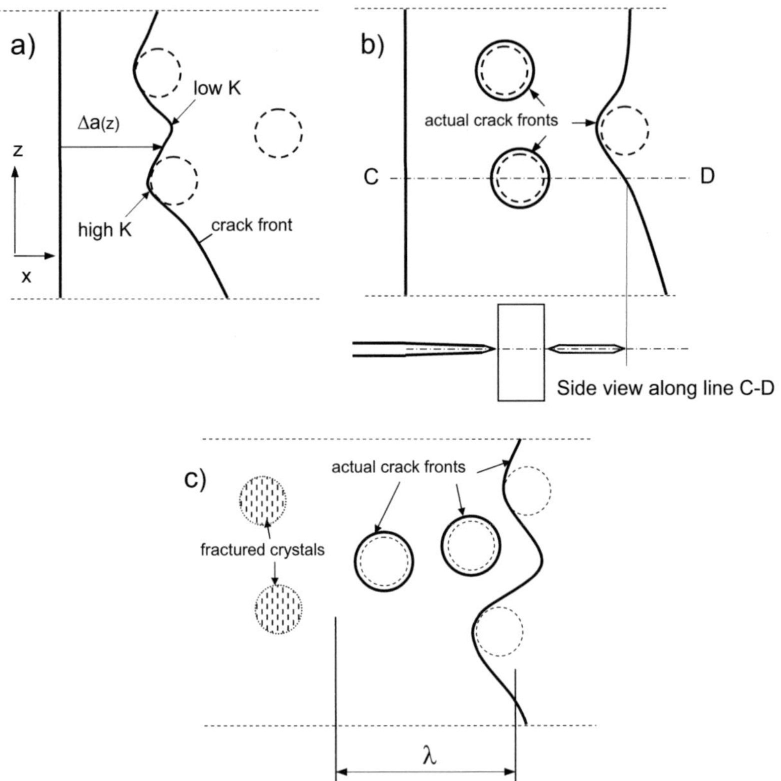

Fig. C3.9 a) Detail of Fig. C3.8b (dashed rectangle) with a continuous crack front, b) crack-front dissolution in case of strongly different crack resistances of matrix and obstacles, c) saturation of the zone containing intact β-crystals behind the main crack front.

When, finally, the applied stress intensity factor at the pinning locations reaches the toughness value $K_{\mathrm{Ic},\beta}$, the remaining β-crystals must fail, too. Consequently, a rather constant steady-state zone of width λ with coexistence of cracked and intact parts must move through the specimen (Fig. C3.9c). The real fracture mechanics problem is even for a simple 2d-crack highly 3-dimensional and cannot be handled by conventional 2d-tools. In order to allow for a simplified treatment the coexistence of cracked and non-cracked parts in this zone may be described by bridging stresses distributed over a zone of length λ.

In this sense, the high bridging stresses for the very short crack-tip distances reflect the fracture toughness of the intact β-crystals behind the main crack front. Since this zone must develop first, an R-curve has to be concluded that initially must rise from K_{I0} to a certain saturation within a crack-length increment of about $\Delta a \approx \lambda$.

The absolute values of such bridging stresses are proportional to the percentage p of non-debonding β-crystals (governed by the critical angle Θ_{cr}).

"True bridging" in the conventional sense extending over clearly longer crack-tip distances and resulting in moderately increasing R-curve contributions need larger crack opening displacements to develop debonding and pull-out effects. These contributions are of course proportional to the percentage of β-crystals ($1-p$) able to debond.

C3.3 Analytical description of bridging laws

C3.3.1 Frictional bridging

The reasons for bridging effects can be frictional and elastic crack-face interlocking. In order to simplify the relations as much as possible it is assumed in the following considerations first that only frictional bridging effects are present.

In the bridging model by Mai and Lawn [C3.18], tractions between the upper and lower crack faces are transmitted by friction. Large grains with the lattice orientation different from the surrounding matrix undergo local residual stresses by thermal mismatch.

In the schematic depiction of Fig. C3.10a, a large grain is shown, acting as a crack-bridging event. The x-component of the thermal mismatch tractions σ_{mis} is indicated.

During crack-face separation resulting in an increasing displacement, δ, a friction stress σ_{fr} occurs which is proportional to the mismatch stress.

The loads transferred by the crack face interactions are localized at single grains. They can be modelled as so-called bridging stresses σ_{br} expressed by

$$\sigma_{br} \cong \begin{cases} \mu\sigma_{mis} & for & \sigma_{mis} < 0 \\ 0 & & else \end{cases} \tag{C3.3.1}$$

defining an *effective* friction coefficient μ.

For frictional bridging stresses, Mai and Lawn [C3.18] described the tractions transferred by a single grain as

$$\sigma_{br} = \begin{cases} \sigma_{max}(1 - \delta/\delta_c) & \text{for} \quad \delta \leq \delta_c \\ 0 & \text{for} \quad \delta > \delta_c \end{cases} \qquad (C3.3.2)$$

illustrated in Fig. C3.10b. The parameter δ_c in (C3.3.2) represents a critical displacement at which the friction disappears, i.e. at which the two crack faces are completely separated. This value was assumed in [C3.19] to depend on the lengths of the grains.

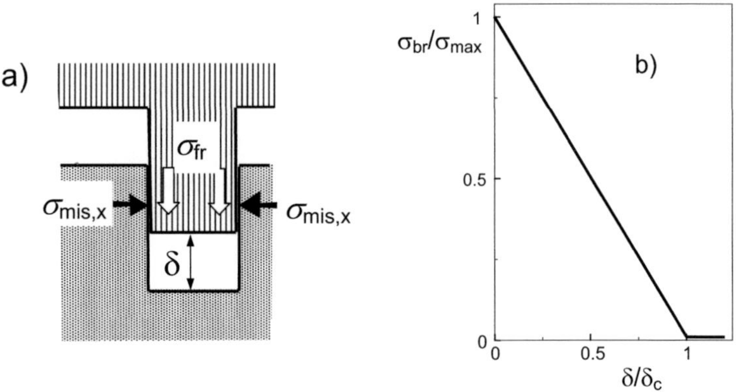

Fig. C3.10 Crack surface interactions due to a local frictional bridging event: a) geometric data, b) transmitted tractions.

From Fig. C1.2a it becomes obvious for the MgY silicon nitride that for the most frequent aspect ratio of ≈ 9, the frequency of grain lengths disappears at disappearing length (length=0) und increases monotonically until reaching the highest frequency at a length of about $1\mu m$. For larger grain lengths, the frequency again decreases monotonically. Such a grain-size distribution and, consequently, the related distribution of the critical displacements δ_c can be described by a Γ-distribution as was proposed in [C3.20]. This distribution expressed by

$$f(\delta_c) = \frac{1}{\delta_{c0}} \left(\frac{\delta_c}{\delta_{c0}} \right) \exp\left(-\frac{\delta_c}{\delta_{c0}} \right) \qquad (C3.3.3)$$

is plotted in Fig. C3.11a.

For the macroscopically averaged bridging stresses it holds

$$\sigma_{br} = \int_0^\infty \sigma_{br,grain} f(\delta_c) d\delta_c \qquad (C3.3.4)$$

and after carrying out the integration

$$\sigma_{br} = \sigma_0 \exp(-\delta / \delta_0)$$
(C3.3.5)

with two material specific parameters σ_0 and δ_0. It should be mentioned that the integration variable (δ_c) not only occurs in the linear function of (C3.3.2) but also in the conditions $\delta \leq \delta_c$ and $\delta > \delta_c$. The global bridging stresses σ_{br} are illustrated in Fig. C3.11b.

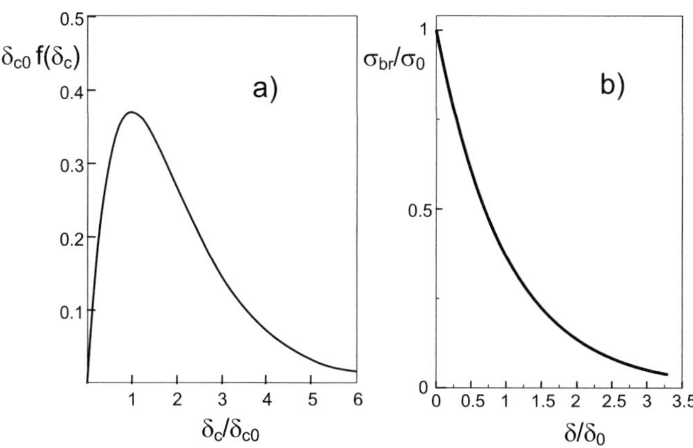

Fig. C3.11 a) Γ-distribution of the critical crack opening displacements, b) global bridging relation for frictional bridging.

C3.3.2 Elastic bridging

An elastic bridging event can result if a crack extends on slightly different planes simultaneously (Fig. C3.12a, see also Fig. A2.3) or by crack extension in the Si_3N_4 with unbroken β-crystals in the crack wake (Fig. C3.12b).

In the case of elastic bridging, the bridging events behave like an elastic spring which fails at a critical displacement δ_c. The "local bridging stress" is [C3.18]

$$\sigma_{br,local} = \begin{cases} \sigma_{max} \, \delta / \delta_c & for \quad \delta \leq \delta_c \\ 0 & for \quad \delta > \delta_c \end{cases}$$
(C3.3.6)

as illustrated in Fig. C3.13a. If it is assumed that also the characteristic displacement δ_c for spring-like bridges is Γ-distributed (i.e. no springs fail without displacement or at an infinite deformation), the "global bridging relation" is given by [C3.20]

$$\sigma_{br} = \sigma_0 \frac{\delta}{\delta_0} \exp(-\delta / \delta_0)$$
(C3.3.7)

with the material specific parameters σ_0 and δ_0. This dependency is shown in Fig. C3.13b. It should be emphasized that the stress parameters σ_{max} and σ_0 are negative. The tractions are against the crack opening displacements.

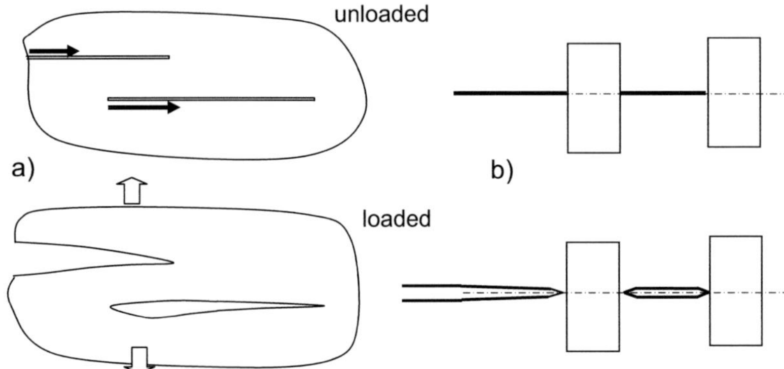

Fig. C3.12 Elastic bridging interaction for a crack in silicon nitride

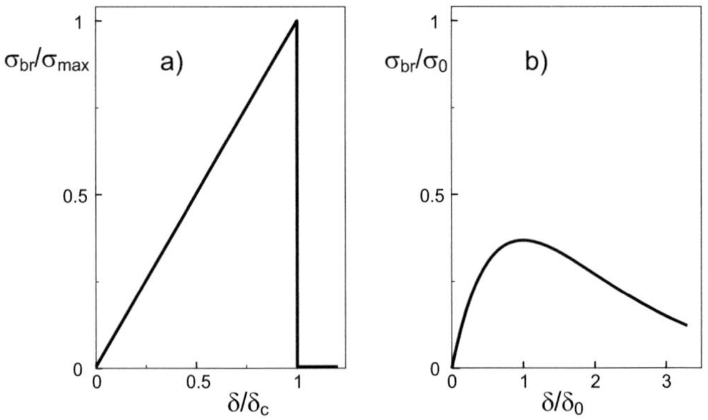

Fig. C3.13 a) Local and b) global bridging relations for elastic bridging.

C3.4 Bridging stresses as a material specific property?

As mentioned before it is the commonly assumed that in the special case of R-curves caused by grain bridging effects, the relation between the bridging stresses and crack opening displacement, $\sigma_{br} = f(\delta)$, is a really intrinsic material property. In this context, we have to make some restrictions. There are measurements available on silicon nitride specimens indicating a possible deviation of bridging stresses obtained from different test specimens exceeding the expected limitations of accuracy.

Figure C3.14 shows as the solid curve the result obtained from crack opening displacement measurements by Kruzic [C3.21] on the side surface of a CT-specimen containing a sharp crack (for the COD results see Section D2.5). The material was MgY-silicon nitride described in Section C1. The dashed curve represents again the result from the R-curve evaluation on notched bars (Fig. C3.5a).

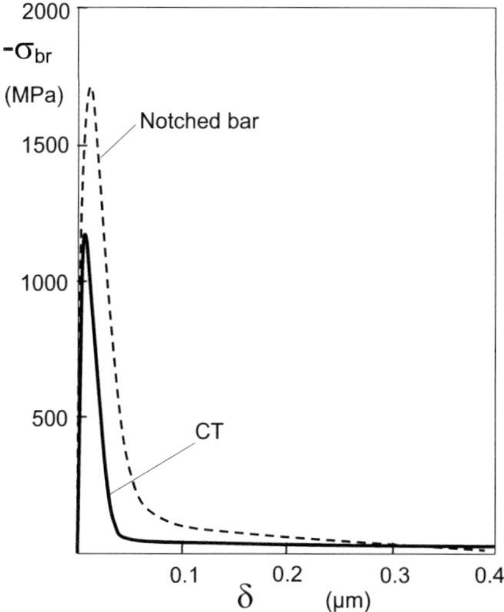

Fig. C3.14 Bridging stresses for the different test specimens.

There are two different measuring conditions, which in principle may be responsible for the deviating bridging stresses. Whereas for the R-curve predominantly the bulk material is responsible, the COD-measurements are mainly affected by the properties of the material at the surface. A second influence may occur due to the strongly different T-stresses in the notched bending bar and the CT-specimen. Both effects will be addressed in this Section.

C3.4.1 Bridging effect expected near free surfaces

In Section C3.2.1, the case of full thermal mismatch stresses was considered as occurring in the bulk of a test specimen. A compressive stress normal to the length axis of a β-crystal will clamp this grain in the matrix material as illustrated in Figs. C3.15a and C3.16a. Grains located near the free surfaces undergo a reduced "clamping" because the stresses perpendicular to this surface must disappear (Figs. C3.15b and C3.16b).

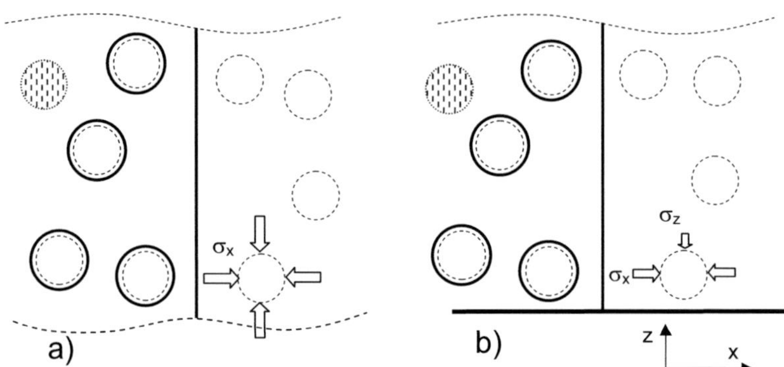

Fig. C3.15 Expected effect of the free surface on thermal mismatch stresses in the crack plane, a) stresses in the bulk material, b) reduction of mismatch stresses near the free surface.

According to the theorem of Saint Venant, the stress reduction will be restricted to a zone of a few grain diameters in thickness and will therefore not contribute noteworthy to R-curves of specimens with several mm thickness. A possible effect can only be expected from direct bridging stress measurements at the surface.

The presence of free side surfaces will have the following consequences on the bridging stresses:

- Reduced compressive stresses normal to the crystal length axis must affect the critical crack terminating angle Θ_{cr} (Fig. C3.16c) with the result of increased debonding lengths ℓ and a reduced percentage p in eq.(C3.2.1) representing the crystals which fail without debonding. Both, the debonded crack and the crack without kinking act as elastic bridges. The debonded cracks act as elastic bridging events with the displacements predominantly caused by the crack-surface displacements and to a nearly negligible extend by the grains. Trivially, the debonded structure is more compliant.

- The percentage of grains, $1-p$, acting as frictional bridges is increased at the surface. On the other hand, this share undergoes reduced friction stresses according to the model by Mai and Lawn [C3.18], Fig. 3.16d. The consequence is a reduced frictional bridging stress intensity factor contribution.

- Bridging stresses due to elastic bridges are reduced. Bridging stresses due to friction are less affected since the increased portion of frictional bridging events is increased but their "strongness" is reduced.

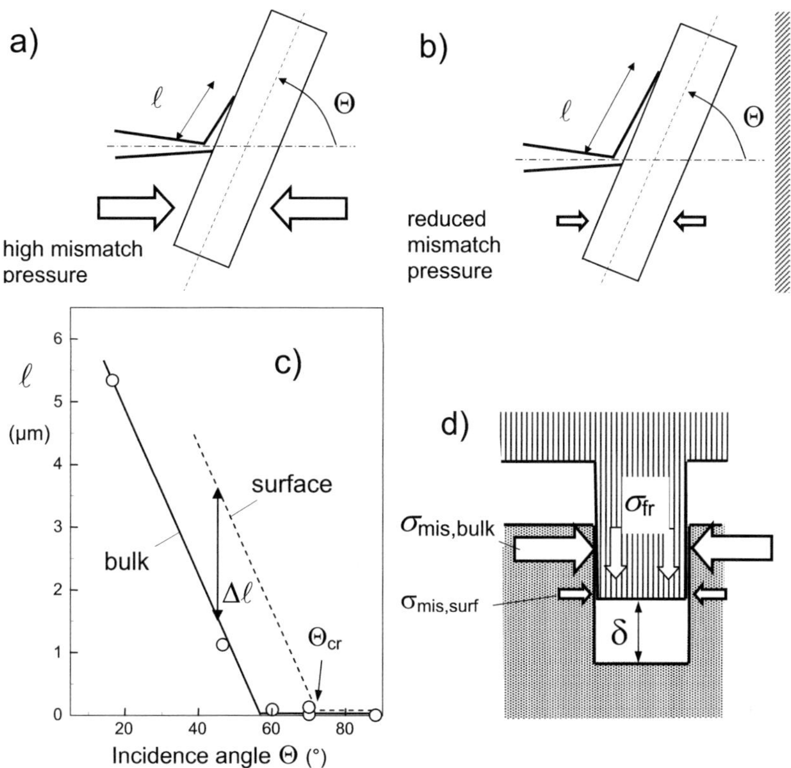

Fig. C3.16 Influence of a reduced compressive stress normal to a β-crystal. Debonding lengths under the same applied load for a) bulk material, b) grains located near a free surface, c) schematic representation of the expected effect according to Fig. C3.7a, d) effect on frictional bridging events.

C3.4.2 Effect of the T-stress component

If K_I denotes the mode-I stress intensity factor, the crack-parallel near-tip stress component σ_x represented by the singular and first regular stress terms (Section B1.1) reads

$$\sigma_x = \frac{K_{tip}}{\sqrt{2\pi r}}\cos\tfrac{1}{2}\varphi(1 - \sin\tfrac{1}{2}\varphi\sin\tfrac{3}{2}\varphi) + T + O(r^{1/2}) \qquad (C3.4.1)$$

where r is the crack-tip distance and φ the polar angle. In the wake of the crack, $\varphi=\pi$, where the bridging interactions are active, only the crack-parallel so called "T-stress" T remains fully present. Its influence may be discussed in a simplified model.

Effect on kinking

Due to the anisotropy and non-homogeneity of a Si_3N_4-ceramic, the quantitative computation of stress intensity factors and T-stresses for the real kink situation of Fig. C3.17a is very complicated. The K-values are now complex numbers. In order to allow a simple understanding of the trends caused by the T-stress, an approximation may be made by using the special Dundurs-parameters of $\alpha=\beta=0$ (assuming unique Young's modulus and Poisson ratio).

The effect of a debond of length ℓ, occurring for an angle of $\Theta<\Theta_{cr}$ is shown in Fig. C3.17a. The crack may terminate in the centre of a β-grain where shear-mismatch disappears. It kinks into the weaker direction of the grain boundary and develops under the additionally acting normal component of the thermal mismatch stress, $\sigma_{mis}<0$, and the T-stress T to a kink length ℓ. The stress intensity factors at the tip of the kink can then be written

$$K_{\mathrm{I}}(\ell) = K_{\mathrm{I}}(a)\cos^3(\tfrac{1}{2}\Theta) + \sigma_{mis}\sqrt{\ell}\sqrt{\frac{8}{\pi}} + T\sqrt{\ell}\sqrt{\frac{8}{\pi}}\sin^2\Theta \qquad (C3.4.2)$$

$$K_{\mathrm{II}}(\ell) = K_{\mathrm{I}}(a)\cos^2(\tfrac{1}{2}\Theta)\sin(\tfrac{1}{2}\Theta) - T\sqrt{\ell}\sqrt{\frac{8}{\pi}}\sin\Theta\cos\Theta \qquad (C3.4.3)$$

where the mode-I stress intensity factor $K_{\mathrm{I}}(a)$ and the T-stress T have to be computed for the straight unkinked crack (i.e. for $\ell\to 0$). The energy release rate for a crack extension by ℓ can be represented by the effective stress intensity factor K_{eff}

$$G = K_{\mathit{eff}}^2/E', \quad E'= E/(1-v^2) \qquad (C3.4.4)$$

$$K_{\mathit{eff}} = \sqrt{K_{\mathrm{I}}(\ell)^2 + K_{\mathit{II}}(\ell)^2} \qquad (C3.4.5)$$

with E=Young's modulus and v=Poisson's ratio. A kink crack in the interface must grow until K_{eff} equals the "fracture toughness" $K_{\mathrm{Ic,int}}$ of the interface. The kink length ℓ then results in terms of the two parameters $\sigma_{mis}/K_{\mathrm{I}}(a)$ and T/σ_{mis} by solving the equation

$$\left(\frac{K_{\mathrm{Ic,int}}}{K_{\mathrm{I}}(a)}\right)^2 = \left(\cos^3(\tfrac{1}{2}\Theta) + \frac{\sigma_{mis}}{K_{\mathrm{I}}(a)}\sqrt{\ell}\sqrt{\frac{8}{\pi}}\left(1+\frac{T}{\sigma_{mis}}\sin^2\Theta\right)\right)^2 +$$

$$+ \left(\cos^2(\tfrac{1}{2}\Theta)\sin(\tfrac{1}{2}\Theta) - \frac{\sigma_{mis}}{K_{\mathrm{I}}(a)}\frac{T}{\sigma_{mis}}\sqrt{\ell}\sqrt{\frac{8}{\pi}}\cos\Theta\sin\Theta\right)^2 \qquad (C3.4.6)$$

For $\ell=0$ the critical incidence angle Θ_{cr} results as

$$\frac{K_{\mathrm{Ic}}}{K_{\mathrm{I}}(a)} = \cos^2(\tfrac{1}{2}\Theta_{cr})$$ (C3.4.7)

For any given incidence angle Θ the kink length ℓ as a function of the incidence angle can be determined from

$$\cos^4(\tfrac{1}{2}\Theta_{cr}) = \left(\cos^3(\tfrac{1}{2}\Theta) + \frac{\sigma_{mis}}{K_{\mathrm{I}}(a)}\sqrt{\ell}\sqrt{\frac{8}{\pi}}\left(1 + \frac{T}{\sigma_{mis}}\sin^2\Theta\right)\right)^2 +$$

$$+ \left(\cos^2(\tfrac{1}{2}\Theta)\sin(\tfrac{1}{2}\Theta) - \frac{\sigma_{mis}}{K_{\mathrm{I}}(a)}\frac{T}{\sigma_{mis}}\sqrt{\ell}\sqrt{\frac{8}{\pi}}\cos\Theta\sin\Theta\right)^2$$ (C3.4.8)

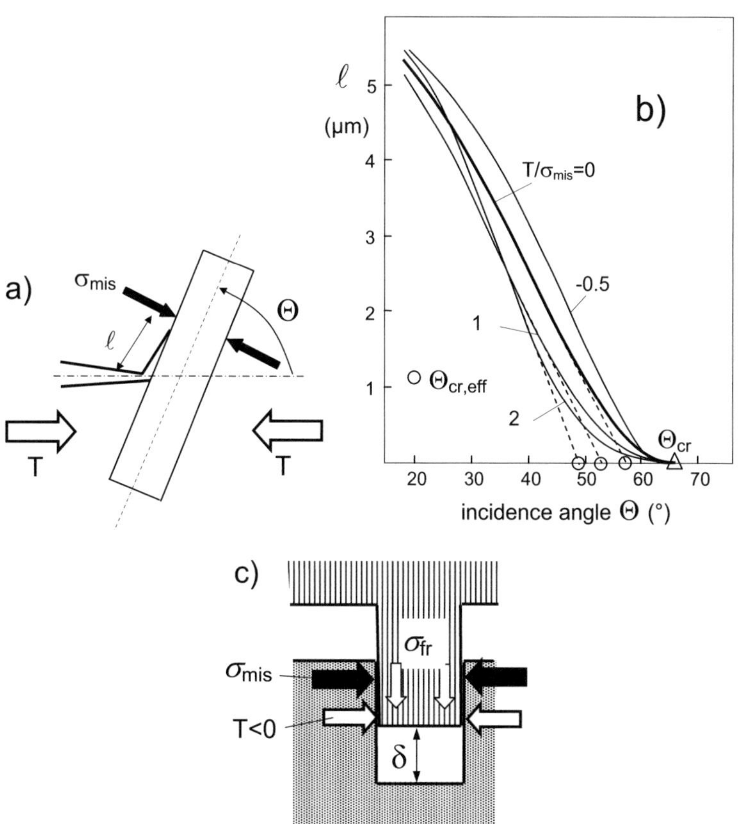

Fig. C3.17 Influence of the T-stress: a) debonding by crack kinking at the interface grain/grain-boundary phase, b) effect of $T<0$ on debonding length, c) on frictional bridging.

For a critical incidence angle of for example $\Theta_{cr}=65°$ (triangle), the kink length as a function of the kink angle is plotted in Fig. C3.17b for the case $T=0$ by the bold line. This result was obtained from eq.(C3.4.8) with a parameter of $\sigma_{mis}/K_I(a)= -73/\sqrt{m}$ that matches the value $\ell=5.3\mu m$ at $\Theta=18°$ as found by Satet [C3.15] for a Si_3N_4-ceramic with La content. The thin lines represent the effect of the T-stress. Since the mismatch stress is negative in Fig. C3.17a, the ratio T/σ_{max} is positive in this case. The linear extrapolation of the lower curve parts to $\ell\rightarrow0$ defines effective critical kink angles indicated by the circles. With increasing T/σ_{max} these effective critical kink angles, designated now $\Theta_{cr,eff}$ decrease.

Since $K_I(a)$ should be in the order of $5<K_R<7$ MPa$\sqrt{m}$ (see Section C2), it results $-\sigma_{mis}\approx350-500$ MPa. It should be noted that rather large T-stress values are necessary for the computed shift of the curves in Fig. C3.17b.

T-stresses in mechanical test specimens are mostly negligible. There are only a small number of tests exhibiting strongly negative T-stresses. Most important are the DCDC-test, the cone crack and, generally, notched specimens showing very small notch radii. It has been shown in [C3.22, C3.23] that for slender notches very strong compressive T-stresses occur in notched bending bars as were used in [C3.6, C3.24] for the R-curve determination (Section B3.2). It can be concluded from [C3.23] that during the first few micrometers of crack extension from a notch strongly negative T-stresses occur that for notch radii of $R<20\mu m$ can reach values in the order up to 50 times the applied bending stress, Fig. C3.18.

For the R-curve experiments on silicon nitride with Y_2O_3 and MgO content, bending bars of $W=4mm$ width were used with starter notches of $a_0\approx2.5mm$ depth having notch radii of $R=6-10\mu m$. The applied bending stresses were about 30-40 MPa. As can be seen from Fig. C3.18, only the first crack extensions of $\Delta a<3\mu m$ are affected by the T-stress term. For this crack extension, a suppression of debonding is partially reached.

From Fig. C3.18 it can be concluded that the T-stresses after about $\Delta a=1\mu m$ crack extension ($T/\sigma_{bend}\cong-18$) are in the order of $T\approx-650$MPa. After $2\mu m$ crack growth ($T/\sigma_{bend}\cong-8$), the T-stress drops to about $T\approx-280$MPa. In terms of Fig. C3.17b, this gives a shift of the effective kink angles from 57° to 51° and 54°, respectively.

Since debonding generally leads to a more compliant structure, the steepness of the R-curve is increased during the very first $2\mu m$ crack extension. An effect worth mentioning is therefore negligible for the representation of $K_R(\Delta a)$-curves for $\Delta a>2\mu m$.

Nevertheless, an effect on the bridging stresses has to be expected. It has been shown in (see [C3.11] and Section C5.2) that the height of the bridging stresses in the crack-tip distance r is in a first-order approximation proportional to the steepness of the R-curve after a crack extension of $\Delta a=r$

$$\sigma_{br}\big|_{r=\Delta a} \propto -\frac{dK_R}{da} \qquad\qquad (C3.4.9)$$

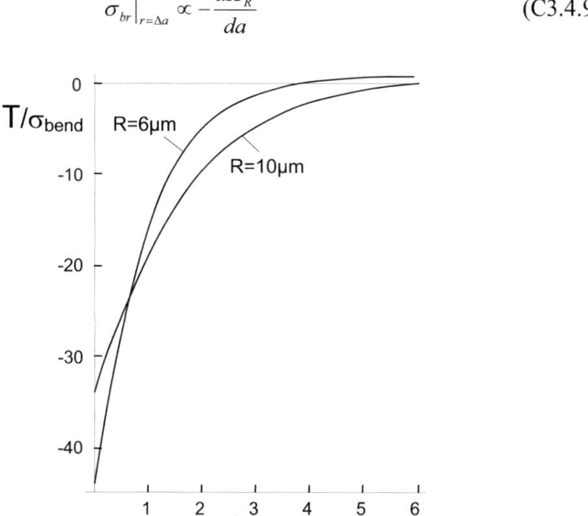

Fig. C3.18 T-stress for small cracks emanating from narrow notches in a bending bar according to [C3.22].

Effect on frictional bridging

Due to the compressive T-stress the friction stress and, consequently, the bridging stresses are increased. From the friction model by Mai and Lawn [C3.18], an amount of $\Delta\sigma_{br}=\mu T$ can be concluded, Fig. C3.17c (see also Section F1).

Final remark:

The measurements on surfaces result in too low bridging stresses. On the other hand the measurements on notched bars exhibit too high values. Therefore, the "true" bridging relation for MgY silicon nitride (Fig. C3.14) should lie between these two limit cases.

To the author's actual knowledge, the bridging stresses relation $\sigma_{br}=f(\delta)$ obtained in the bulk of a test specimen under the condition $T=0$ should represent the true material property.

References C3

C3.1 Pezzotti, G., Muraki, N., Maeda, N., Satou, K., Nishida, T., In situ measurement of bridging stresses in toughened silicon nitride using Raman microprobe spectroscopy, J. Am. Ceram. Soc. **82**(1999), 1249-56

C3.2 Pezzotti, G., Ichimaru, H., Ferroni, L., Raman Microprobe Evaluation of Bridging Stresses in Highly Anisotropic Silicon Nitride, JACS **84**(2001), 1785-90.

C3.3 Kruzic, J.J., Cannon, R.M., Ager III, J.W., Ritchie, R.O., Fatigue threshold R-curves for predicting reliability of ceramics under cyclic loading, Acta Mater. **53**(2005), 2595-2605.

C3.4 Fett, T., Munz, D., Kounga Njiwa, A.B., Rödel, J., Quinn, G.D. Bridging stresses in sintered reaction-bonded Si_3N_4 from COD measurements, J. Europ. Ceram. Soc. **25**(2005), 29-36.

C3.5 Fett, T., Munz, D., Thun, G., Bahr, H.A., Evaluation of bridging parameters in Al2O3 from R-curves by use of the fracture mechanical weight function, J. Amer. Ceram. Soc. **78**(1995) 949-51.

C3.6 Fett, T., Fünfschilling, S., Hoffmann, M.J., Oberacker, R., Jelitto, H., Schneider, G.A., R-curve determination for the initial stage of crack extension in Si_3N_4, J. Am. Ceram. Soc. **91**(2008), 3638-42.

C3.7 Fett, T., Munz, D., Influence of narrow starter notches on the initial crack growth resistance curve of ceramics, Arch. Appl. Mech. (2006), DOI 10.1007/s00419-006-0055-3.

C3.8 Fett, T., Munz, D., Stress Intensity Factors and Weight Functions, Computational Mechanics Publications, (1997) Southampton, UK.

C3.9 Fünfschilling, S., Fett, T., Hoffmann, M.J., Oberacker, R., Jelitto, H., Schneider, G.A., Determination of the crack-tip toughness in silicon nitride ceramics, J. Mater. Sci., **44**(2009), 335-338.

C3.10 S. Fünfschilling, T. Fett, R. Oberacker, M.J. Hoffmann, G.A. Schneider, P.F. Becher, J.J. Kruzic, Crack-tip toughness from Vickers crack-tip opening displacements for materials with strongly rising R-curves, J. Am. Cer. Soc., **94**(2011), 1884–1892.

C3.11 Fünfschilling, S., Fett, T., Gallops, S.E., Kruzic, J.J., Oberacker, R., Hoffmann, M.J., First- and second-order approaches for the direct determination of bridging stresses from R-curves, J. Eur. Ceram. Soc. **30**(2010), 1229-1236.

C3.12 Fünfschilling, S., Influence of micro-structure on the R-curve behaviour of silicon nitride ceramics (in German), Thesis, IKM 53, KIT Scientific Publishing, Karlsruhe, 2009; (Open Access: http: /creativecommons.org.lizenses/by-nc-nd/3.0/de/).

C3.13 Fünfschilling, S., Fett, T., Hoffmann, M. J., Oberacker, R., Schwind, T., Wippler, J., Böhlke, T., Özcoban, H., Schneider, G.A., Becher, P.F., Kruzic, J.J., Mechanisms of toughening in silicon nitrides: The roles of crack bridging and microstructure, Acta Materialia **59**(2011), 3978-89.

C3.14 Becher, P.F., Hsueh, C.H., Angelini, P.,Tiegs, T.N., Toughening behavior in whisker-reinforced ceramic matrix composites, J. Am. Ceram. Soc. **71**(1988), 1050-1061.

C3.15 Satet, R., Einfluß der Grenzflächeneigenschaften auf die Gefügeausbildung und das mechanische Verhalten von Siliciumnitrid-Keramiken, Thesis, IKM-Report 038, April 2003, Universität Karlsruhe.

C3.16 Gao, H., Rice, J.R., A first order perturbation analysis of trapping by arrays of obstacles, ASME, J. Appl. Mech. **56**(1989), 828-36.

C3.17 Roux, S., Vandembroucq, D., Hild, F., Effective toughness of heterogeneous brittle materials, Eur. J. of Mech. A /Solids **22**(2003), 743-749.

C3.18 Mai, Y., Lawn, B.R., Crack-interface grain bridging as a fracture resistance mechanism in ceramics: II. Theoretical fracture mechanics model, J. Am. Ceram. Soc. **70**(1987), 289.

C3.19 Fett, T., Munz, D., Bridging stress relations for ceramic materials, Europ. Ceram. Soc. **15**(1995) 377-83.

C3.20 T. Fett, D. Munz, Evaluation of R-curve effects in ceramics, J. Mater. Sci. **28**(1993) 742-52.

C3.21 Kruzic, J.J., unpublished results.

C3.22 Fett T. Stress intensity factors, T-stresses, Weight function, IKM50, Universitätsverlag Karlsruhe, Karlsruhe; 2008.

C3.23 Fett, T., Munz, D., Influence of narrow starter notches on the initial crack growth resistance curve of ceramics, Arch. Appl. Mech. **76**(2006), 667-679.

C3.24 Fünfschilling, S; Fett, T; Oberacker, R; Hoffmann, MJ; Oezcoban, H; Jelitto, H; Schneider, GA; Kruzic, JJ, R-curves from compliance and optical crack-length measurements, J. Am. Ceram. Soc., **93**(2010), 2814-21.

C4 R-curve and bridging stresses for SiAlONs

C4.1 R-curves of SiAlONs

The same tests and evaluation procedures as outlined in detail in Section C2 for hot-pressed silicon nitrides were also performed for the SiAlON-ceramics described in Section C1 [C4.1]. Figure C4.1a shows a part of the load-displacement curve near the load maximum. The load vs. compliance change ΔC is represented in Fig. C4.1b.

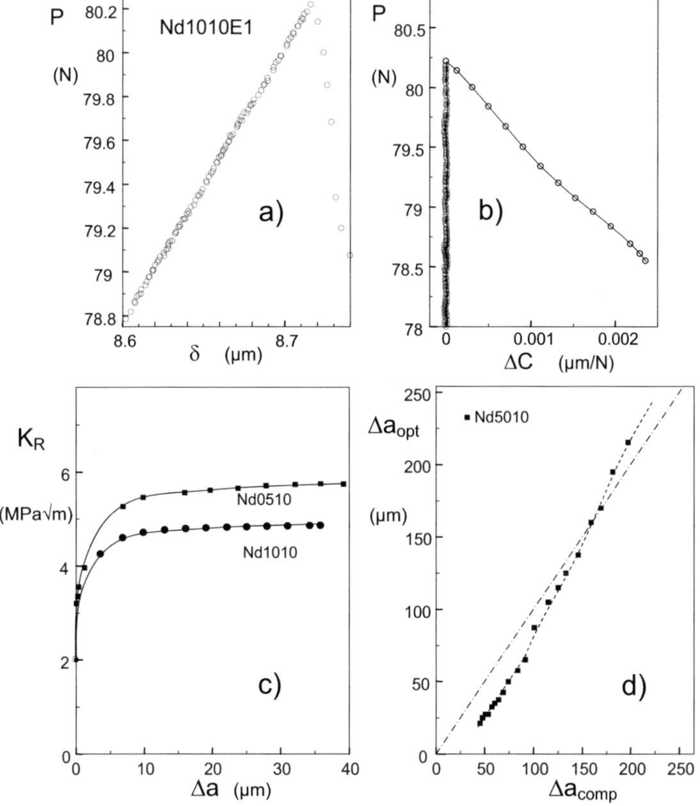

Fig. C4.1 a) load vs. displacement curve for a SiAlON from Riva [C4.1], b) related curve with the increase of compliance as the abscissa, c) R-curves for two ceramics, d) relation between optically observed crack length vs. result by compliance evaluation.

The R-curves in Fig. C4.1c show similar behaviour as the results obtained for the silicon nitrides. Already after a crack-length extension of abut $\Delta a=10\mu m$, the R-curve has nearly reached saturation. Also the relation between crack lengths from optical and compliance evaluation is similar, Fig. C4.1d.

C4.2 Bridging stresses for SiAlONs

The evaluation of bridging stresses according to Section C3 yields the stress distributions given in Fig. C4.2a. The near-tip displacement field is shown in Fig. C4.2b. Also for the SiAlON ceramics the early deviation from a simple $\delta\propto\sqrt{r}$ relation is obvious. The bridging stress relation $\sigma_{br}(\delta)$obtained by combining the results of Fig. C4.2a and C4.2b is shown in Fig. C4.3.

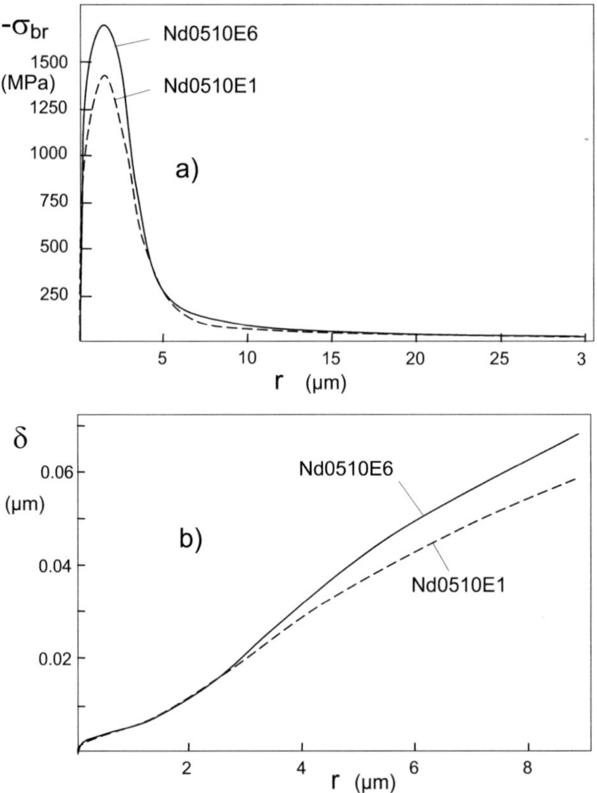

Fig. C4.2 a) Distribution of the bridging stresses, b) crack opening displacements near the crack tip.

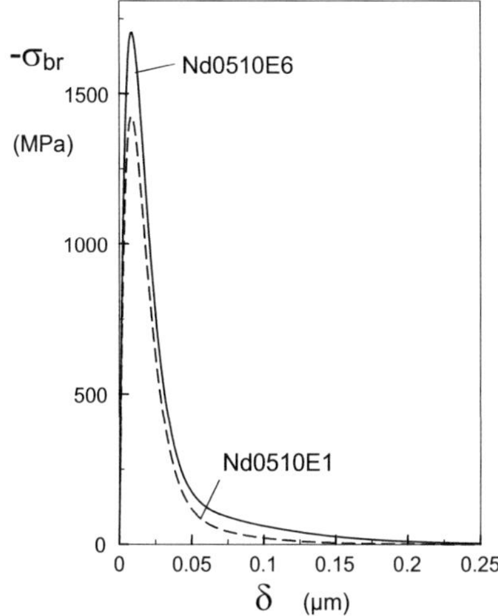

Fig. C4.3 Bridging laws for the SiAlON-ceramics of Fig. C4.2.

C4.3 Crack-tip toughness K_{I0}

Crack opening displacement measurements on Vickers indentation cracks in SiAlON ceramics produced with an indentation load of 98.8 N were performed by Riva [C4.1]. Here only the results from Fig. C4.4 may be reported (for details of the procedure see Section D2).

In Fig. C4.4a the crack opening displacements (COD) versus crack-tip distance are plotted. Figure C4.4b represents the measured COD plotted versus the theoretical COD, δ_{comp}, computed for an applied stress intensity factor of $K=1$ MPa$\sqrt{m}$ according to Section B2.2. The slope of the near-tip straight-line gives the crack-tip stress intensity factor K_{I0}. The region of evaluated data was up to $x \approx 7$ µm from the tip. The single results for the 3 materials are compiled in Table C4.1 with an average of $K_{I0} \cong 1.5$ MPa$\sqrt{m}$ obtained by the common evaluation of all data points. The crack-tip toughness for the SiAlONs is below that of the Si_3N_4 (Section D2).

SiAlON	K_{I0} (MPa$\sqrt{m}$)	90% CI
Nd0510E1-3	1.56	[1.50, 1.63]
Nd0510E6-1	1.33	[1.28, 1.39]
Nd1010E1-1	1.58	[1.50, 1.66]
All data combined	1.51	[1.47, 1.55]

Table C4.1 Evaluation of crack-tip toughness for the investigated materials; data in brackets 90% confidence intervals (CI).

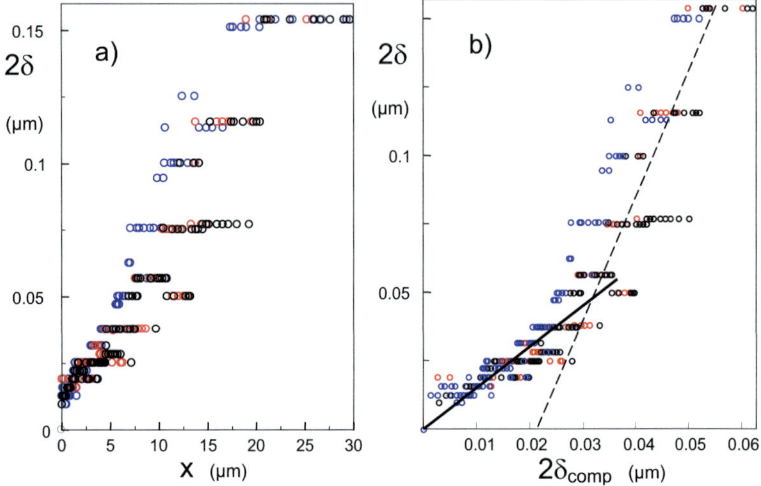

Fig. C4.4 Crack-opening from Vickers indentations for the SiAlON-ceramics by Riva [C4.1]; a) Representation of the crack opening displacements for the 3 materials, b) measured displacements plotted versus the displacements computed for $K=1$ MPa√m, straight-line fit for all data of part (b) with the fitting range determined according to [C4.2], solid line: near-tip displacements, dashed line: far-tip displacements (symbols: black Nd0510E6, blue Nd0510E1, red Nd1010E1).

References C4

C4.1   Riva, M., Entwicklung und Charakterisierung von Sialon- Keramiken und Sialon-SiC-Verbunden für den Einsatz in tribologisch hochbeanspruchten Gleitsystemen, IKM 54, KIT Scientific Publishing, Karlsruhe, 2011.
C4.2   S. Fünfschilling, T. Fett, R. Oberacker, M.J. Hoffmann, G.A. Schneider, P.F. Becher, J.J. Kruzic, Crack-tip toughness from Vickers crack-tip opening displacements for materials with strongly rising R-curves, J. Am. Cer. Soc., **94**(2011), 1884–1892.

C5 Approximate procedures

C5.1 Errors by application of linear-elastic compliance

The compliance method for an evaluation of crack length including the feedback of the bridging stresses on the compliance was discussed in Section C2.1.2. Direct measurement of the crack length by using a microscope showed deviating results. The optically obtained crack length increments were in general smaller than those found via compliance. The effect of the different crack length increments $\Delta a^{(el)}$ from the elastic compliance and the physical crack length Δa_{phys} on the R-curve data is illustrated in Fig. C5.1a. The solid circle may represent an arbitrarily chosen individual data point $K_R(\Delta a^{(el)})$ computed with the crack length $\Delta a^{(el)}$ as obtained by application of the linear-elastic compliance relation eq.(B1.5.2).

As an example, we now assume that the physically "true" crack length increment Δa_{phys} might be 50% larger. This assumption results first in a shift of the data point to a larger Δa and second the increased crack length increment results trivially in a larger K-value. The new data point $\{\Delta a_{phys}, K_R(\Delta a_{phys})\}$ is indicated by the open circle. Although the location of this data point in Fig. C5.1a is clearly changed, a deviation of this point from the given R-curve is hardly ascertainable, and the R-curve remains essentially unchanged. This holds for all the investigated curves on Si_3N_4 but cannot be incorporated as a general result valid for other materials.

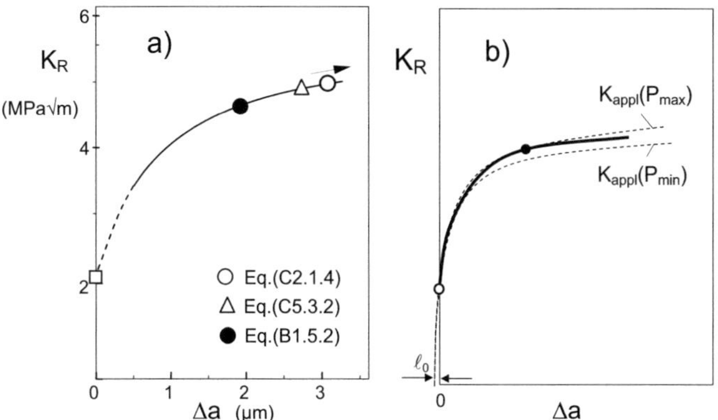

Fig. C5.1 a) Influence of an error in crack length on the steep R-curves for cracks ahead of slender notches in silicon nitride (for the triangle see Section C5.3), b) schematic plot of a steep R-curve (solid curve) and the minimum and maximum K_{appl} curves (dashed lines) in an R-curve test.

The reason for this slightly astonishing result can easily be understood. In the experiments, crack extension (indicated by an increasing compliance) occurs under very small load variations of clearly less than 1% as can be seen from Fig. C2.1 for material SL200BG. During the displacement-controlled test it always holds $K_{appl}=K_R$.

The minimum load in Fig. C2.1 is $P_{min}=70.4$ N, the maximum load $P_{max}=70.7$ N. For small crack extensions, the maximum and minimum limits for the applied stress intensity factor result from (B3.2.1) and (B3.2.2) as

$$K_{appl}(P_{min})\tanh(D\sqrt{\ell_0+\Delta a}) \le K_R \le K_{appl}(P_{max})\tanh(D\sqrt{\ell_0+\Delta a}) \qquad (C5.1.1)$$

where the notch geometry enters via $D \approx 2.243/\sqrt{R}$. The limit curves are indicated in the schematic plot of Fig. C5.1b as the dashed curves. It should be noted that in this plot the variations in load and, consequently, the differences in $K_{appl}(P_{min})$ and $K_{appl}(P_{max})$ are strongly exaggerated.

From eq.(C5.1.1) it becomes clear that for any crack length Δa, correct or slightly incorrect, the data pair $(K_R, \Delta a)$ will be located on nearly the same unchanged curve at least in the range in which the applied loads are sufficiently constant.

For materials with steeply rising R-curves, including many Si_3N_4 ceramics and the sialon ceramics as well, only small errors, approximately 1%, will be incurred using the simple traction-free linear-elastic compliance evaluation that accounts for the notch effect. Thus, using that method is preferable for saving time and effort in the data evaluation. However, such a procedure will give significant errors at larger crack extensions because then the applied load significantly varies. However, at longer crack extensions optical crack length measurements converge with the actual crack length, and thus can be used to reliably evaluate the crack length.

C5.2 Procedure for estimating bridging stresses from R-curves

C5.2.1 First-order solution

It has to be emphasized that the procedures for solving the system of eqs.(C3.1.2) and (C3.1.4) are rather complicated and need much numerical effort. This was the reason for the development of first- and second-order approximations [C5.1, C5.2] which were successfully applied to silicon nitride [C5.3] and alumina ceramics.

Using the weight function representation, the bridging stress intensity factor can be expressed by the distribution of bridging stresses σ_{br} acting in the wake of the crack

$$K_{br} = \int_0^a h(r,a)\,\sigma_{br}(r)\,dr \qquad (C5.2.1)$$

with the total crack length a consisting of the initial crack length a_0 free of bridging, and the crack extension Δa showing bridging, i.e. $a=a_0+\Delta a$.

In the following derivation of the bridging stresses from (C5.2.1), the special shape of the bridging stress distribution

$$\sigma_{br} = \begin{cases} f(r) & for \quad r \le \Delta a \\ 0 & for \quad r > \Delta a \end{cases} \tag{C5.2.2}$$

is used implicitly. This relation allows splitting of the integral in two parts

$$K_{br} = \int_0^{\Delta a} h(r,a)\sigma_{br}(r)\,dr + \underbrace{\int_{\Delta a}^a h(r,a)\sigma_{br}(r)\,dr}_{=0} \tag{C5.2.3}$$

The first-order procedure consists in the following steps:

By taking the derivative with respect to a on both sides of (C5.2.1) it results

$$\frac{dK_{br}}{da} = -\frac{dK_R}{da} = h(r,a)\sigma_{br}(r,a)\Big|_{r=\Delta a} + \int_0^a \frac{\partial h(r,a)}{\partial a}\sigma_{br}(r,a)\,dr$$

$$\tag{C5.2.4}$$

$$+ \int_0^a h(r,a)\frac{\partial \sigma_{br}(r,a)}{\partial a}\,dr$$

(It should be mentioned that in the related equations in [C5.2] the lower integration limit a_0 has to be replaced by "0", see erratum to [C5.2] in [C5.^{4}]).

The bridging stresses result from (C5.2.4) as

$$\sigma_{br}\Big|_{r=\Delta a} = -\frac{1}{h\big|_{r=\Delta a}}\frac{dK_R}{da} - \frac{1}{h\big|_{r=\Delta a}}\int_0^a \frac{\partial h(r,a)}{\partial a}\sigma_{br}(r,a,\Delta a)\,dr$$

$$\tag{C5.2.5}$$

$$- \frac{1}{h\big|_{r=\Delta a}}\int_0^a h(r,a)\frac{\partial \sigma_{br}(r,a,\Delta a)}{\partial a}\,dr$$

Identifying $\sigma_{br}(r)=\sigma_{br}\big|_{r=\Delta a}$ gives a *first-order solution* for the bridging stresses:

$$\sigma_{br}^{(1)}(r) = -\frac{1}{h}\Big|_{\Delta a=r}\frac{dK_R}{da}\Big|_{\Delta a=r} \tag{C5.2.6}$$

The subscript $\Delta a = r$ in (C5.2.6) means: For the computation of the bridging stresses in a certain crack-tip distance r, e.g. $r=5$ µm, the derivative of the R-curve after $\Delta a=5$ µm has to be introduced as well as the value of the weight function h in distance $r=5$ µm.

C5.2.2 Second-order solution

In the first-order approximation, only the first term of eq. (C5.2.5) was regarded with the bridging stresses depending exclusively on r. In a second-order solution, the first integral term of (C5.2.5) accounting for the influence of crack length via the weight function $h(r,a)$ may be included. In this approximation, the first-order solution (C5.2.6) has to be used under the integral, resulting in

$$\sigma_{br}^{(2,1)}(r) = \sigma_{br}^{(1)}(r) - \frac{1}{h\big|_{r=\Delta a}} \int_0^a \frac{\partial h(r',a)}{\partial a} \sigma_{br}^{(1)}(r')\, dr'\bigg|_{\Delta a=r} \tag{C5.2.7}$$

where the first number in the superscript counts for the second term of (C5.2.5), and the second number for the step of iteration (started with $m=1$). Generally it holds for higher iteration numbers $m>1$

$$\sigma_{br}^{(2,m)}(r) = \sigma_{br}^{(1)}(r) - \frac{1}{h\big|_{r=\Delta a}} \int_0^a \frac{\partial h(r',a)}{\partial a} \sigma_{br}^{(2,m-1)}(r')\, dr'\bigg|_{\Delta a=r} \tag{C5.2.8}$$

C5.2.3 Example of application

An R-curve for Mg-La-containing silicon nitride is shown in Fig. C2.2a. The crack propagation test was carried out on a pre-notched bending bar. The best fit of the measured data was given by

$$K_R = K_{I0} + C_1(1-(1+C_2\sqrt{\Delta a})\exp[-C_2\sqrt{\Delta a}]) \tag{C5.2.9}$$

with the "best" set of coefficients compiled in Table C2.1.

For the determination of the bridging stresses, the weight function for a small crack ahead of a slender notch has to be applied. It is given by eqs.(B3.2.6) and (B3.2.9). Figure B3.2 shows this solution together with the geometric data R, ℓ, and r. For very short crack extensions in the order of about $\ell \leq R$, the weight function significantly depends on ℓ/R. Application of eq.(C5.2.6) yields the first-order bridging stresses represented in Fig. C5.2a by the dash-dotted curve. The first and second iterative solutions of eq.(C5.2.8) are shown by the thin continuous curves. The third and fourth iterations showed differences less than 2 MPa, therefore, they are represented by the same dashed curve. This curve indicates the final curve of convergence, i.e. the bridging stress distribution $\sigma_{br}(r)$.

The full solution obtained by solving the simultaneous system of integral equations numerically, eqs.(C3.1.2) and (C3.1.4), and fitted by

$$\sigma_{br} \cong \sum_{n=0}^{N} \sigma_n \frac{\delta}{\delta_n} \exp[-\delta/\delta_n] \qquad\qquad (C5.2.10)$$

with $N=1$, $\sigma_0 = -3375$ MPa, $\delta_0=0.0118\mu m$, $\sigma_1 = -450$ MPa, $\delta_1=0.0476\mu m$, is shown in Fig. C5.2b as the solid curve. Although differences to the approximations occur, the second-order solution clearly exhibits all characteristic features of the exact solution, which are in this case:

a) Very high bridging stresses of $\sigma_{br} \approx -1250$ MPa,

b) a strong concentration of the bridging effects in a crack tip distance of about $r=0\text{-}10\mu m$.

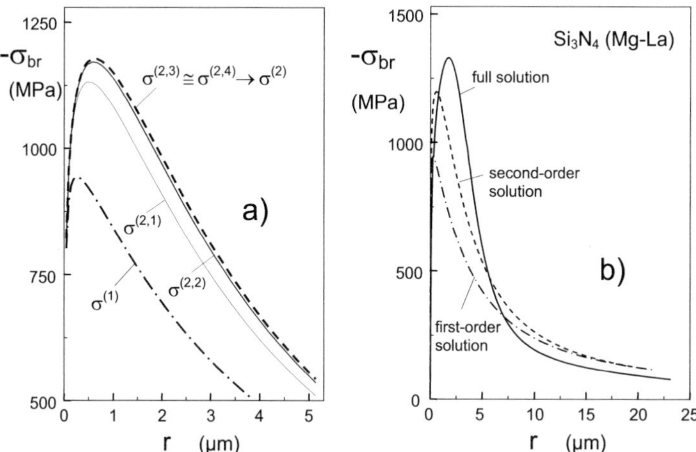

Fig. C5.2 a) Convergence study of the approximations; dash-dotted curve: first-order solution from (C5.2.6), continuous curves: first and second iteration steps for the second-order solution, dashed curve: identical third and fourth iteration steps indicate the final second-order solution, b) first- and second-order bridging stresses compared with the full solution obtained by simultaneous solving the integral equations (C3.1.2) and (C3.1.4).

The reason for the differences between the approximations and the full solution becomes evident by the representations in Fig. C5.3. Figure C5.3a shows the total crack profiles for a long crack extension of $\Delta a=50\mu m$ as the thick curve. For shorter crack-growth phases of $\Delta a=2\mu m$, $5\mu m$, $10\mu m$ and $20\mu m$, the total displacements were additionally computed via eq.(C3.1.4). The results are given by the thin curves with the circles indicating the individual crack length increments.

For a very short crack extension of Δa=2µm, the total displacements differ maximum by about 37% from the long-crack displacements. Consequently, also the bridging stresses must differ for such a small amount of crack propagation.

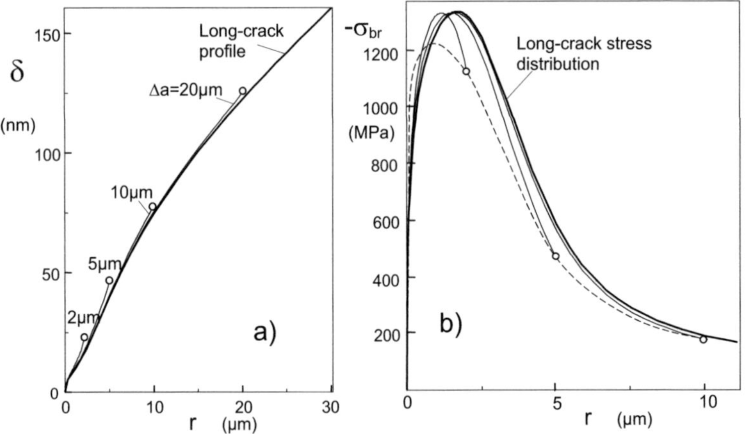

Fig. C5.3 a) Total displacement profiles for short crack extension phases, b) related deviations of the bridging stress distributions, dashed curve tentatively introduced as the interpolation of the stresses at $r=\Delta a$, represented by the circles.

The related bridging stress distributions are given in Fig. C5.3b exhibiting clearly the differences in the stresses. These differences cause the different stress distributions for the full solution and the second-order approximation in Fig. C5.2b. In this context it should be taken into account that the stress values resulting from eqs.(C5.2.6) and (C5.2.8) are those acting at $r=\Delta a$, i.e. at the end of the bridging zone. These values are indicated by the circles in Fig. C5.3b. The dashed curve, which interconnects these data points, exhibits the main features of the second-order solution as slightly lower stress level and peak stress at a shorter crack-tip distance.

Approximate results $\sigma_{br}^{(2,\infty)}$ according to eq.(C5.2.8) are compiled in Fig. C5.4a for several of the investigated materials. These solutions show the same ranking in maximum bridging stresses as the full solutions shown in Fig. C3.4. A comparison of the bridging laws resulting from the two attempts is given in Fig. C5.4b for the Mg-Y-containing silicon nitride. The approximate solution according to eq. (C5.2.8) and the full solution yield bridging stresses only slightly different in shape but with roughly the same area under the σ_{br} vs. δ curve, i.e. the same crack-face separation energy.

Finally, it should be emphasized that the effort in time was for the approximate procedure only about 1% of that for the full evaluation.

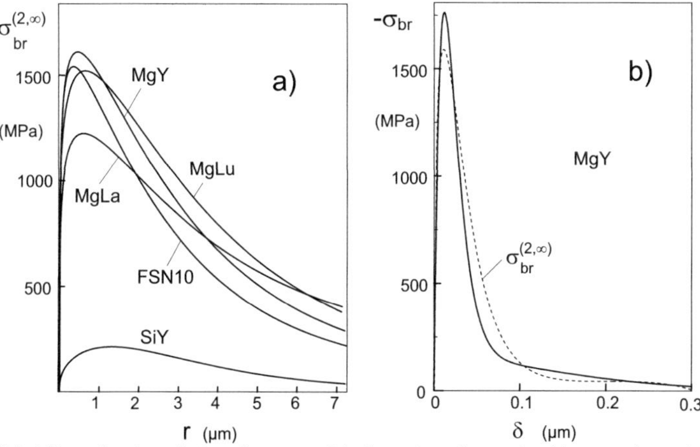

Fig. C5.4 a) Second-order solutions for some of the investigated ceramics, b) comparison of approximate (dashed curve) and full solution (bold curve) for MgY.

C5.3 Approximate solution for the bridging effect on the compliance

The determination of the crack length from the bridging-affected compliance was described in Section C2.1.2. The bridging stresses had to be computed from a system of integral equations (C2.1.6)-(C2.1.10). Instead of the very time consuming "full" solution, the first-order solution for the bridging stresses σ_{br} in the crack wake may be applied according to eq.(C5.2.6) that reads

$$\sigma_{br}\big|_{r=\Delta a} \cong -\frac{1}{h}\frac{dK_R}{da}\bigg|_{\Delta a=r} \tag{C5.3.1}$$

The change of the total displacements by action of externally applied tractions and intrinsic bridging stresses simultaneously, are for 4-point bending given in terms of the related compliances

$$\Delta C \cong \Delta C_{appl}(\Delta a) - \frac{2B}{P_{appl}^2}\int_0^{\Delta a}\frac{\delta_{appl}}{h}\frac{dK_R}{da}\bigg|_{\Delta a=r} dr \tag{C5.3.2}$$

with the measured compliance ΔC, the compliance in the absence of bridging interactions, ΔC_{appl}, and the contribution by bridging interactions expressed by the integral term. The displacements of crack faces due to the applied load, δ_{appl}, are given by eq.(C2.1.10). The unknown crack length increment Δa occurs at two places in (C5.3.2) explicitly, namely in the argument of the compliance function for the applied load and at the upper integration limit. Implicitly, the crack length also affects the K_R-term of the integrand. This fact calls for an iterative solution using the method of "successive approximation". For its explanation eq.(C5.3.2) may be slightly re-written as

$$\Delta C \cong \Delta C_{appl}(\Delta a^{(n+1)}) - \frac{2B}{P_{appl}^2} \int_0^{\Delta a^{(n+1)}} \left(\left. \frac{\delta_{appl}}{h} \frac{dK_R}{da} \right|_{\Delta a=r} \right)^{(n)} dr \qquad (C5.3.3)$$

In the first approximation step, the compliance according to Section B1.5 is used as a first-order solution to compute the crack depth $a^{(1)}$. From this, the first-order approximations for the applied displacements, the weight function and the R-curve K_R have to be computed as indicated by the superscript (n), here $n=1$. Using the first-order terms under the integral, eq.(C5.3.3) can be solved by application of a zero routine providing the second-order approximation $a^{(2)}$. The procedure has to be repeated until a certain degree of convergence is reached. This establishes the final crack depth increment Δa. Although this procedure is clearly shorter than the full treatment of Section C2.1.2, computer time may still need several hours.

The evaluation via (C5.3.2) is shown in Fig. C5.1a by the triangle which is in sufficient agreement with the exact evaluation via eq.(C2.1.4).

References C5

C5.1 Fünfschilling, S., Mikrostrukturelle Einflüsse auf das R-Kurvenverhalten bei Siliciumnitridkeramiken, Thesis, IKM 53, Universitätsverlag Karlsruhe, 2009.
C5.2 Fünfschilling, S., Fett, T., Gallops, S.E., Kruzic, J.J., Oberacker, R., Hoffmann, M.J., First- and second-order approaches for the direct determination of bridging stresses from R-curves, J. Eur. Ceram. Soc. **30**(2010), 1229-1236.
C5.3 Fünfschilling, S., Determination of R-curves from corrected load-displacement data and the calculation of the bridging stresses from the R-curves, 6th International Conference on Nitrides and Related Materials (ISNT 2009), 15.-18.03.2009, Karlsruhe.
C5.4 Corrigendum to: Fünfschilling, S., Fett, T., Gallops, S.E., Kruzic, J.J., Oberacker, R., Hoffmann, M.J., First- and second-order approaches for the direct determination of bridging stresses from R-curves, J. Eur. Ceram. Soc. 31(2011), 1523.

D1 Effect of crack and specimen type

D1.1 R-curves for 1-d cracks

Knowledge of the bridging relations as the real material property allows to compute the R-curves for any crack and loading case by application of eqs.(C3.1.2) and (C3.1.4) if the weight function solution for the special crack is available. Some results may be given here for NC132 and the MgY-ceramic using the data of Table C3.1. The measured R-curve from a test on an edge-notched bending bar of NC132 with relative notch depth a_0/W=0.75 and notch-root radius R=20µm is represented by the solid curve in Fig. D1.1a. The weight function for this configuration was taken from [D1.1]. In addition the R-curve for a compact-tension (CT)-specimen with a sharp initial crack of a_0/W=0.25 was computed also with the weight function from [D1.1]. The result is shown by the dashed curve. This curve is about 0.1-0.15 MPa√m below that for the notched bending bar. Figure D1.1b shows three R-curves for bending bars. The solid curve again represents the result for the deep notch with a_0/W=0.75. The slightly deviating dashed curve holds for a sharp crack of same depth. A notch of only a_0/W=0.25 yields the dash-dotted curve. Figure D1.1c shows the R-curve for MgY (solid curve) together with the computed R-curve (dashed curve) for tests with the double cleavage drilled compression (DCDC) specimen using the bridging stresses from eq.(C3.1.9) and the weight function (B1.3.12-B1.3.14). There is only a small effect of the different DCDC-weight function visible.

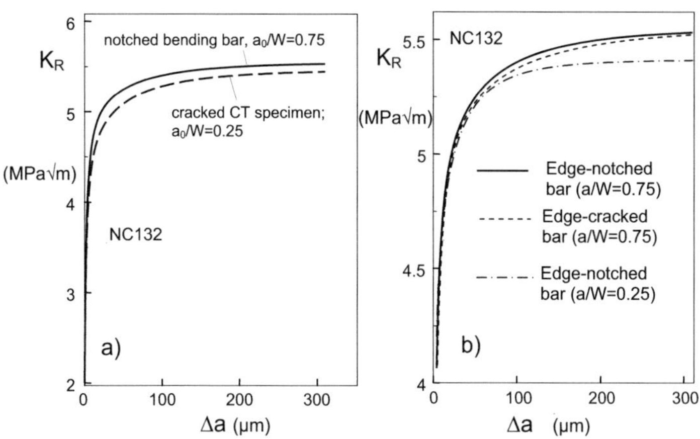

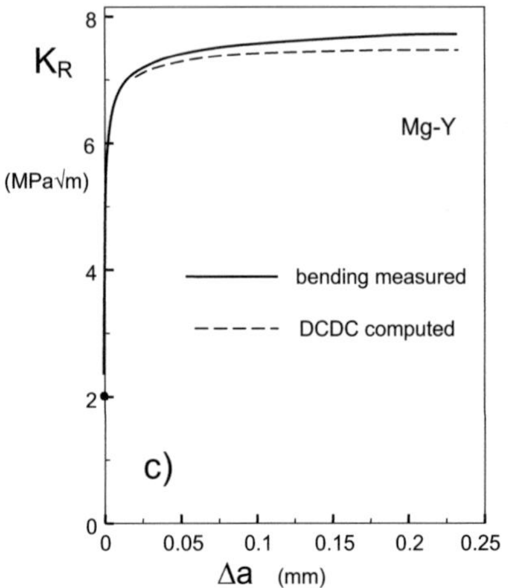

Fig. D1.1 a) and b) Influence of different specimens on the R-curve, demonstrated for NC132, c) for DCDC-test on MgY.

The different R-curves of Fig. D1.1a illustrate the influence of the different weight functions for bars and CT-specimens and Fig. D1.1b the effect of different crack or notch depths. The main reason for the reduced K_R is the lower weight function which for the short notch with $a_0/W=0.25$ is hardly influenced by the back wall effect in contrast to the deep notch or crack (see Fig. D1.2).

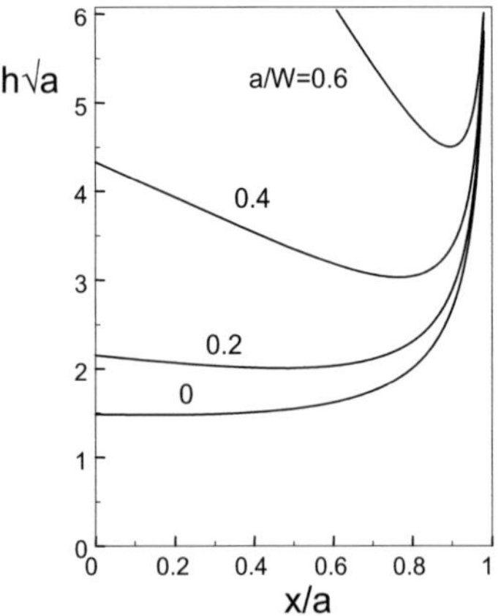

Fig. D1.2 Influence of the relative crack depth a/W on the weight function for an edge-cracked bending bar of width W (x=crack-mouth distance).

D1.2 R-curves for 2-d cracks

D1.2.1 Pure tension and wedging

The R-curve for a semi-circular surface crack under remote tension loading can be computed with the weight function method according to eqs.(B2.1.1) and (B2.1.3). The result is plotted in Fig. D1.3a. This curve is slightly below the R-curve for the edge-notched bending bar with maximum deviations of 0.15 MPa√m. The differences are caused by the different weight functions as is obvious from the comparison in Fig. D1.4.

In addition an indentation crack with an impression radius of 50% of the crack radius was evaluated with the weight function taken from [D1.2]. This crack exhibits a crack-growth resistance reduced by 0.25-0.4 MPa√m. Here the strongest difference of the weight function with respect to the edge-notched bar has to be taken into account as was outlined in detail in [D1.3].

There are two further reasons for the slightly lower K_R–curve of the Vickers indentation crack compared with the semi-circular surface crack, namely:

(1) The bridged area is smaller for indentation cracks due to the existence of the closed compressive zone without stressed bridges.

(2) The bridging stresses and, consequently, the R-curve of the indentation crack are additionally reduced due to the large crack opening displacements near the indenter impression (Fig. D1.5).

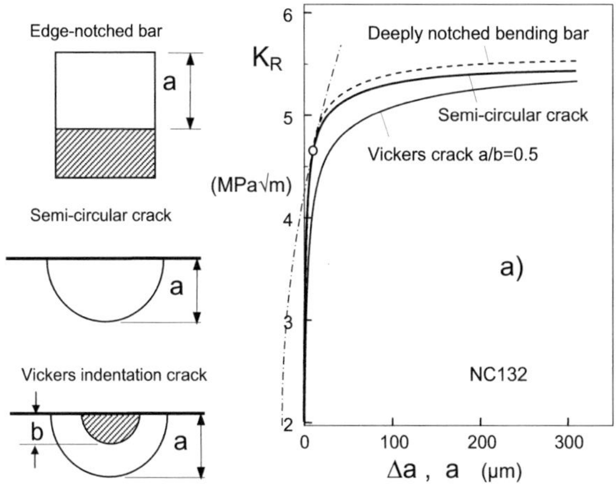

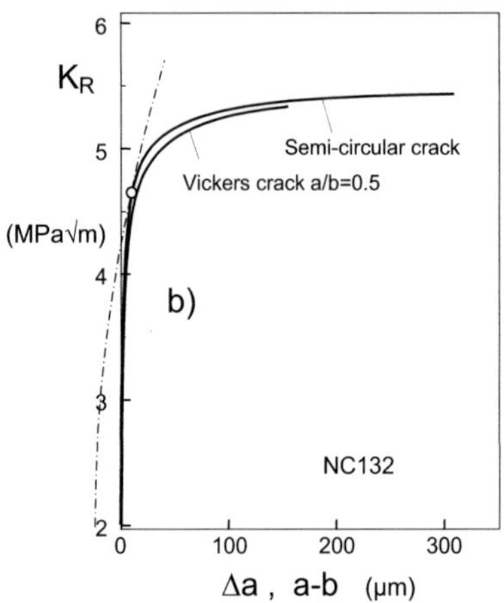

Fig. D1.3 a) R-curves for a semi-circular surface crack under external tensile load and Vickers indentation crack loaded by the residual stress field extending over half of the crack radius (dashed curve
for the deeply notched bending bar for comparison). Abscissa Δa for notched bar and semi-circular
crack, a for Vickers indentation crack, b) representation of the R-curves with the abscissa for the indentation crack changed to a-b, dash-dotted curve: K_{appl} for a semi-circular crack with depth of 25 µm,
circle: K_{Ic} from the tangent condition.

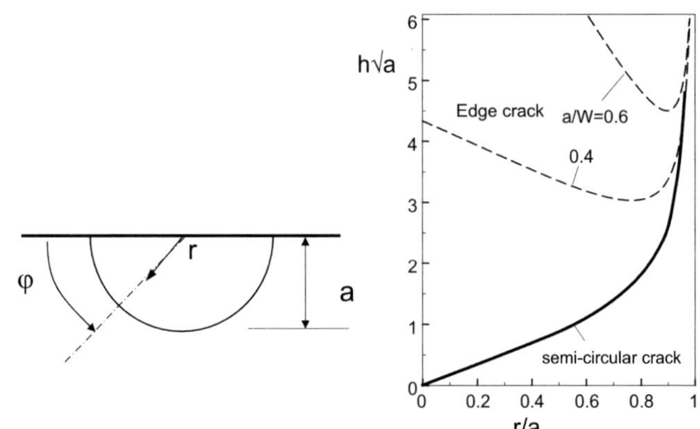

Fig. D1.4 Weight function for a semi-circular crack (solid curve); radial distance from the surface=r
(edge crack included for comparison).

The dash-dotted curve in Fig. D1.3a and D1.3b shows the K_{appl}-curve for a semi-
circular surface crack with a depth of 25 µm causing failure at the tangent point (circle) to the R-curve. The stress intensity factor at this point is called the fracture tough-

ness K_{Ic}. In the case of the 25µm deep crack, one obtains $K_{Ic}\cong4.6$ MPa$\sqrt{}$m. This value depends on the initial crack length. For shorter cracks, a lower K_{Ic}-value is obtained.

Figure D1.3b shows the R-curve for the indentation crack plotted versus the open part of the crack undergoing bridging stresses, i.e. versus a-b. This transformation eliminates the effect of point (1) mentioned before. The remaining small difference reflects the effect of the large crack opening displacements of the Vickers indentation crack illustrated schematically in Fig. D1.5.

Figure D1.3b gives rise for a rather rough approximation of the R-curve for an indentation crack for the case that only the R-curve for a semi-elliptical crack under tension is available. It then holds

$$K_{\text{Vickers}}(a) \approx K_{\text{semi-circle}}(\Delta a = a - b) \qquad\qquad (\text{D1.2.1})$$

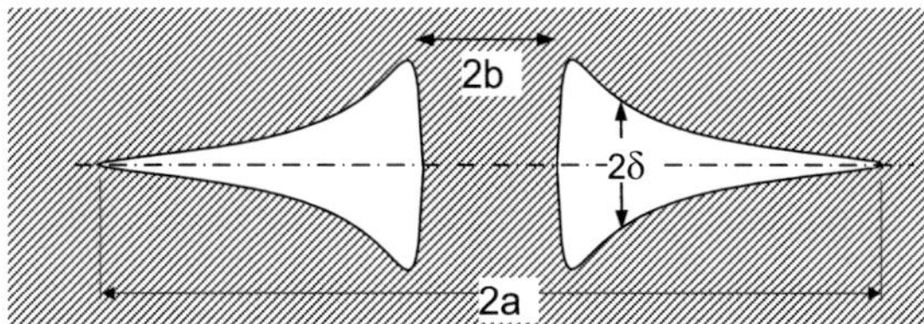

Fig. D1.5. Crack profile at the free surface of a surface crack (top view) under a residual stress field according to Hill [D1.4]; schematic COD distribution with displacements exaggerated.

Asymptotic solution for bridging stresses with small ranges

For very steep R-curves, i.e. bridging stresses with a very small range, the bridging effect is mainly concentrated in the near-tip region. Instead of the complicated displacement fields, it is possible to consider the near-tip solution of δ exclusively. The displacements δ are then replaced by the near-tip displacement δ_{tip} described by the Irwin parabola

$$\delta_{tip} = \sqrt{\frac{8}{\pi}}\,\frac{K_{I0}}{E}\,\sqrt{a-r} \qquad\qquad (\text{D1.2.2})$$

which must hold for any type of crack if $a\text{-}r \ll a$.

If the bridging law is for instance of the type (C3.3.7), the distribution of bridging stresses can be estimated without solving a complicated integral equation system. This results in

$$\sigma_{br}(r) \approx \sigma_0 \sqrt{\frac{8}{\pi}\frac{K_{I0}}{\delta_0 E}}\sqrt{a-r}\,\exp\left[-\sqrt{\frac{8}{\pi}\frac{K_{I0}}{\delta_0 E}}\sqrt{a-r}\,\right] \qquad (D1.2.3)$$

By introducing this stress distribution into

$$K_{br} = \int_{a-b}^{a} h_{semi-circ}\,\sigma_{br}(r)\,dr, \qquad (D1.2.4)$$

the approximate bridging stress intensity factors and the approximate R-curves for small crack extensions can be determined by simple integration.

D1.2.2 Superimposed tension and wedging

In Fig. D1.3, the cracks either were loaded by remote tension or by wedging due to the residual stresses in the damaged zone beneath the indenter. In many practical applications, these two types of loading have to be superimposed. This is for instance the case for an indentation crack, which after removal of the indenter is loaded by an externally applied load in order to perform a strength test.

For the total load consisting of the residual and the external part, an R-curve can be constructed by simple interpolation because the differences of the two curves are small.

If the crack radius after the indentation test is a_0 and after application of an additional external load $a=a_0+\Delta a$, the R-curve may be written

$$K_R = \frac{a_0}{a}K_{Vickers} + \frac{\Delta a}{a}K_{semi-circle} \qquad (D1.2.5)$$

This relation trivially fulfils the limit cases of $K_R=K_{Vickers}$ for $a=a_0$ and $K_R \rightarrow K_{semi}$ for $a>>a_0$.

References D1

D1.1 Fett, T., Munz, D., Stress Intensity Factors and Weight Functions, Computational Mechanics Publications, (1997) Southampton, UK.

D1.2 Fett T. Stress intensity factors, T-stresses, Weight function, Supplement Volume IKM55, KIT Scientific Publishing, Karlsruhe; 2009; (open Access: http://digbib.ubka.uni-karlsruhe. de/ volltexte/ 1000013835).

D1.3 Fett, T., Fünfschilling, S., Hoffmann, M. J., Oberacker, R., Different R-curves for 2- and 3-dimensional cracks, Int. J. Fract. **153**(2008), 153-159.

D1.4 Hill, R., Mathematical Theory of Plasticity, *Oxford University Press*, Oxford (1950).

D2 Results from Vickers indentation cracks

D2.1 Crack tip stress intensity from Vickers COD measurements

In R-curve experiments on silicon nitride ceramics (see Section C2) it was found that about 90% of the final crack resistance is already reached after Δa=10μm. Furthermore, the determination of accurate K_{I0} values from such R-curve data is essentially impossible. Since the range of Δa in which K_R rises from K_{I0} to the saturation value is extremely short, an extrapolation of the measured R-curve to $\Delta a = 0$ is simply not accurate.

An independent method to measure K_{I0} is the evaluation of crack-opening displacements (COD). This procedure has been widely used in literature by evaluation of cracks in the scanning electron microscope [D2.1-D2.5] and the atomic force microscope (AFM) [D2.6-D2.8]. Using Vickers indentations for this procedure has the advantage that the maximum crack-tip loading is reached after removing the indenter so the COD measurements can be made on unloaded samples. In contrast, the use of straight-through cracks in fracture mechanics test specimens (bending bars [D2.9], CT-specimens, etc.) needs much more effort in keeping the load applied constant while making measurements in the microscope.

For materials with a flat crack resistance curve, like glass, K_{I0}-values can be simply calculated from COD measurements via the so-called Irwin parabola (Fig. D2.1a). Conversely, COD measurements for materials with a rising R-curve must be affected by the bridging stresses. This fact is especially problematic for materials with very steep R-curves. Maximum stress levels are present at a crack-tip distance of only $\approx$0.5-2μm. These high stresses must cause deviations from the Irwin parabola even very close to the crack tip. The consequences of this fact will be outlined in this Section.

Since the final extension of the semi-elliptical "median-radial" Vickers cracks develops during the unloading phase, these cracks are characterized by the condition of $K_I = K_{I0}$ at the crack tip after load removal assuming additional lateral or cone cracks do not also form and relax the residual stress field. Figure D2.2a illustrates the impression of the indenter with diagonal $2b$ and the cruciform semi-circular crack system with diameter $2a$. The total mode-I crack opening $2\delta_I$ normal to the x-axis can be measured under the scanning electron microscope (SEM).

An analytical solution for the COD of Vickers indentation cracks in the absence of bridging stresses was given in [D2.10] and can be described by the relations in Section B2.2.

The first term in eq.(B2.2.8) represents the behaviour very close to the crack tip, called generally the Irwin parabola:

$$\delta_I = \sqrt{\frac{8}{\pi}}\,\frac{K_I}{E}\,\sqrt{x}\,.$$

(D2.1.1)

The relations (B2.2.4) and (B2.2.8) have been found to be effective for determining the crack-tip stress intensity for Vickers indents in soda-lime glass [D2.10]. In Fig. D2.1a, SEM crack opening displacement measurements on a crack introduced in soda-lime glass under 50 N load are plotted as circles. The curve in Fig. D2.1a results from a least squares fit of the measured data in a modified representation. For this purpose the measured crack opening displacements, $\delta_{I,meas}$, are plotted versus the displacements, $\delta_{I,comp}$, computed with eq.(B2.2.4) for a stress intensity factor of $K_I = 1\,\mathrm{MPa\sqrt{m}}$. Because of eq.(B2.2.4), the related plot must then yield a straight line through the origin. Figure D2.1b shows the measured data in this modified representation.

A least-squares fit of the linear dependency yields K_I as the slope of the straight line, in the present example resulting in $K_I = 0.38\,\mathrm{MPa\sqrt{m}}$. Use of this value then yields the solid curve introduced in Fig. D2.1a.

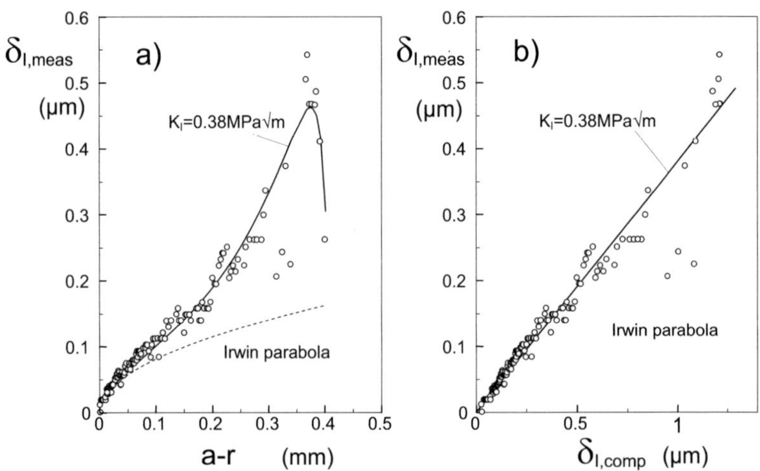

Fig. D2.1 Determination of $K_{I,tip}$ for a soda-lime glass, a) measured crack opening displacement, b) measured COD plotted versus computed COD (K_I=1 MPa√m), from [D2.10].

The dashed line shown in Fig. D2.1a corresponds to the Irwin solution for the near-tip displacement field, eq.(D2.1.1), at the same stress intensity factor. Thus, the COD method can determine the crack tip K_I value in non-bridging ceramics based on the slope of the $\delta_{I,meas}$ - $\delta_{I,comp}(K_I = 1\,MPa\sqrt{m})$ curve. This method was expanded to apply for materials with steeply rising R-curves [D2.11].

D2.2 Vickers COD-curve for a material with a steep R-curve

The test procedure according to [D2.11] may be explained on material MgY (see Section C1) with the R-curve rising to about $K_R \approx 7\,MPa\sqrt{m}$ even after 10 μm crack propagation. For the determination of K_{I0}, Vickers indentation tests were performed on polished surfaces with an indentation load of 98.8 N and a dwell time of 15s. Before the indentation tests, the surfaces were polished through diamond slurries of decreasing diamond grit sizes down to 1μm, and then slightly plasma etched for 1min with a 2:4 CF_4:O_2 gas mixture.

Figure D2.2a shows a schematic top view of the total crack and D2.2b a SEM image of a Vickers indentation crack in very close to the tip. As a consequence of eq.(D2.1.1) valid for any crack shape, the mode-I stress intensity factor K_I and consequently also the special value K_{I0} are related to the mode-I displacements δ_I, exclusively. Therefore, only the displacements normal to the x-axis have to be taken into account for the determination of K_{I0}.

The crack path in a poly-crystalline material is not a very straight line where the orientation of the crack-plane is always in agreement with the x-axis. The needed value δ_I is always larger than the actual normal-distance δ_n of the two opposite crack flanks oriented under an angle φ to the x-direction (see the lower part of Fig. D2.2b). These two measures are identical only if the case of crack segments aligned exactly in x-direction. The difference can become very large in segments deviating strongly from the parallel case. An evaluation along the cross section C-C' would result in roughly $\delta_n \approx \frac{1}{2}\delta$.

Considering this fact, the δ_I-evaluation can most effectively be done for crack parts which are roughly in agreement with the x-axis, i.e. for small angles φ.

On a micro-structural scale, the visible deviations of the local crack normal from the y-axis give rise for a mixed-mode displacement δ_{II} and a mode-II stress intensity factor K_{II}. Since on the macroscopic scale a large number of positive and negative crack-normal deviations occur, the total (average) mode-II displacements and stress intensity factor must disappear too. With other words, the average data curve indicates the global crack behaviour with $K_{II,total}=0$. The scatter in measurements has to be related at least partially to the local mode-II stress intensity factors.

Results for the silicon nitride with Y_2O_3 and MgO content are given in Fig. D2.3a and D2.3b (circles) as a function of the distance x from the tips for 5 individual cracks. There is a rather high data scatter visible. In Section D2.4, the scatter behaviour is discussed based on a large number of measurements for two additional silicon nitrides.

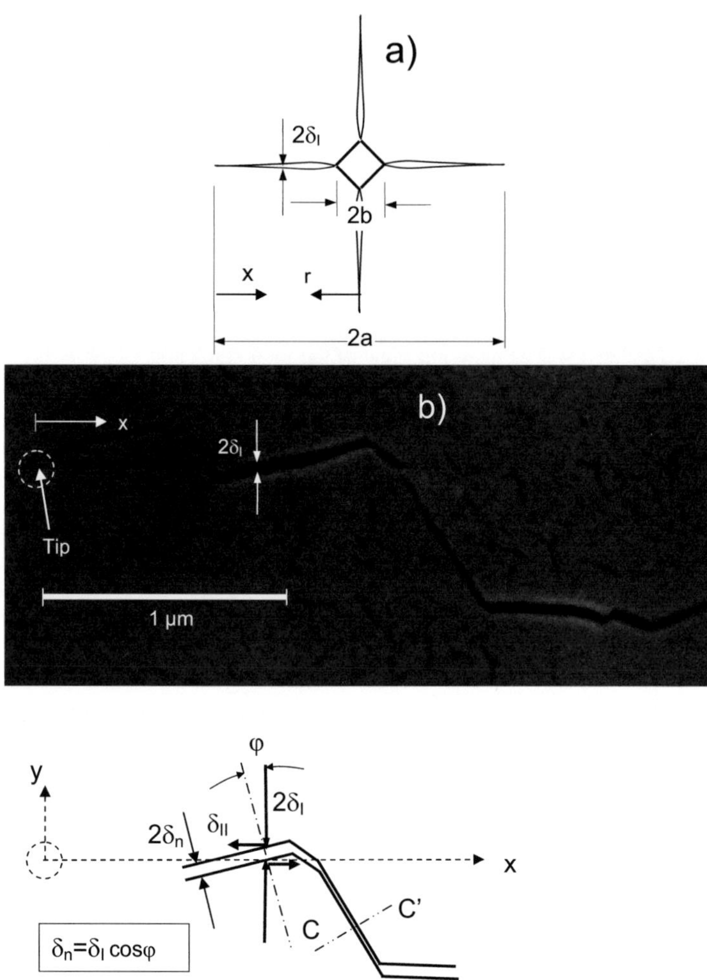

Fig. D2.2 a) Top view on the Vickers indentation cracks, b) crack opening in the near-tip region for a Vickers indentation crack introduced in a non-etched surface with COD-measurement normal to the global crack plane.

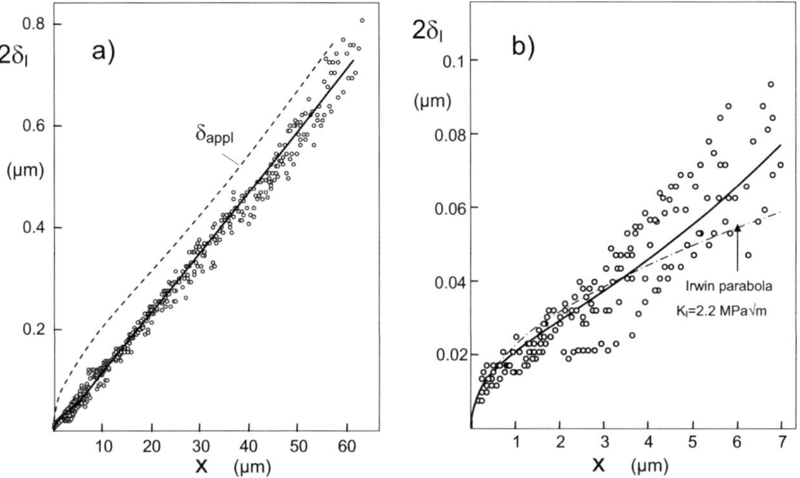

Fig. D2.3 a) Measured COD of Vickers indentation cracks in Si_3N_4 with Y_2O_3 and MgO, b) near-tip results of a).

D2.3 Computation of the full COD-curve

Exact solutions for the crack opening displacements are given only for the case of an internal penny-shaped crack. In the case of a semi-circular crack, the bridging displacements, δ_{br}, caused by the bridging stresses, σ_{br}, result according to Section B2.2

$$\delta_{br}(r) \cong \frac{4}{\pi E'}\int_r^a \left(\int_0^{a'} \frac{r'(1+c(1-r'/a'))\sigma_{br}(r')}{\sqrt{a'^2-r'^2}}\,dr' \right) \frac{(1+c(1-r/a'))da'}{\sqrt{a'^2-r^2}} = \delta - \delta_{appl} \qquad (D2.3.1)$$

where δ stands for the measurable total displacements normal to the global crack plane, i.e. for δ_l in Fig. D2.2b (for $r=a-x$, see Fig. D2.2a). The quantity δ_{appl} is the "applied displacement" caused by the residual stress field in the central region of diameter $2b$. The applied displacements can be computed via eqs.(B2.2.4) or (B2.2.8).

The solution of eq.(D2.3.1) is explained in Section D3 in detail together with a procedure that allows the bridging stresses to be determined. Here only the result for the displacements may be quoted.

The best approximation to the measured data is shown in Fig. D2.3a by the solid curve. Figure D2.3b gives the result in the crack-tip region.

Figure D2.3b also contains the Irwin parabola as the dash-dotted curve for $K_{I0} = 2.20$ MPa√m. This curve now results as:

$$\delta_{tip} = \sqrt{\frac{8}{\pi}}\,\frac{K_{I0}}{E}\,\sqrt{x}\,.$$

(D2.3.2)

It must be noted that the above procedure required much effort and called for an approximation.

D2.4 Near-tip CODs and scatter behaviour

As can be detected already from Fig. D2.3b, there are clear differences between the total near tip displacements and the Irwin parabola both representing the same K_{I0}. Therefore, the two curves are plotted again in Fig. D2.4a.

The two curves agree for very small crack-tip distance, as theory requires. Tolerating errors less than 10% we can conclude that for the MgY-Si_3N_4 material the application of the Irwin parabola is restricted to distances of $x < 0.6\ \mu m$. In the region $0.6\ \mu m < x < 4\ \mu m$ the Irwin parabola overestimates the total displacements and for all larger distances, $x > 4\ \mu m$, underestimates the total COD.

Consequently, the simplified evaluation of K_{I0} by using the Irwin parabola must be restricted to very small crack-tip distances with the disadvantages that

a) only a few data points can be included in the evaluation,

b) the data points near the tip are of lowest accuracy because the displacements, as well as the distances x, show largest percentage errors.

Difficulties in fitting only the near crack tip data can be seen in Fig. D2.4b. To overcome the disadvantages and to avoid the very high numerical effort for the solution of the integral equations, a modified COD-evaluation was proposed in [D2.11].

The principal influence of bridging stresses on the crack opening displacement profiles may be shown for a bridging law according to eq.(C3.1.9) consisting only of the first term:

$$\sigma_{br} = \sigma_0\,\frac{\delta}{\delta_0}\,\exp\left[-\frac{\delta}{\delta_0}\right].$$

(D2.4.1)

In order to study the influence of the bridging range, the two parameters δ_0 and σ_0 were varied simultaneously so that the magnitude of the asymptotically reached bridging stress intensity factor, K_{br}, remains the same. This condition is trivially fulfilled for a constant area under the curve $\sigma_{br}(\delta_l)$, i.e. for a separation energy $\sigma_0\delta_0$=constant, since:

$$\int_0^\infty \sigma_{br}\,d\delta = \sigma_0\delta_0\,.$$

(D2.4.2)

For the following numerical computations K_{I0}=2 MPa√m and $\sigma_0\,\delta_0$=3×10^{-5} MPa m were chosen, which resulted in $K_{br}\rightarrow -2.7$ MPa√m and $K_R\rightarrow 4.7$ MPa√m.

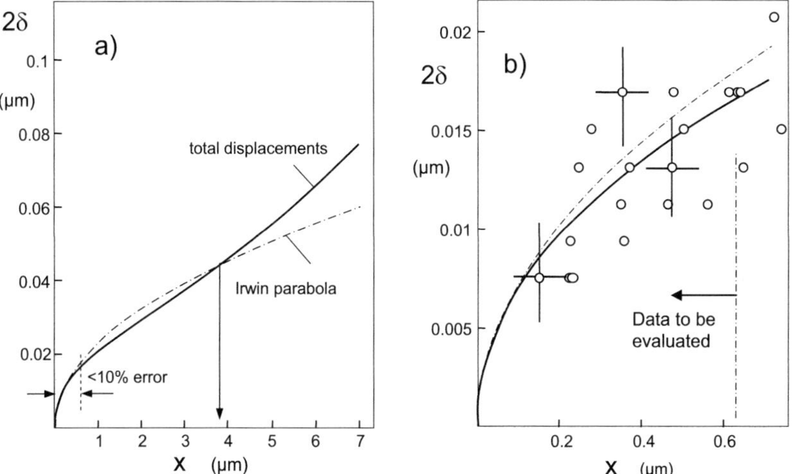

Fig. D2.4 a) Comparison of the total near-tip displacements (solid curve) with the Irwin parabola, b) examples of highly scattered data for the MgY-Si$_3$N$_4$ material (vertical bars indicate a resolution of ±5nm, horizontal bars: resolution for the crack-tip location: expected to be ±50nm).

Figure D2.5a shows the results obtained for the parameter set σ_0=3000 MPa, δ_0=10^{-8} m. The solid continuous curve gives the total crack opening displacement δ. The distribution of the bridging stresses is represented by the dotted curve. The displacement field for a material without a bridging effect (i.e. $K=K_{I0}$) is given by the thin continuous line and the Irwin parabola by the dashed curve. In Fig. D2.5b, the displacements normalized by the bridging range, δ_0, are plotted versus the displacements obtained for a constant applied stress intensity factor of K=1 MPa√m also normalized by δ_0. In this representation the curves for the different parameters practically coincide at least for δ/δ_0<3.

The curves in Fig. D2.5b show that for δ/δ_0<0.1 the slopes of the near-tip curves with and without bridging agree. The slopes are in both cases 2, i.e. K_{I0}=2 MPa√m, as was chosen for the computations. The curve for σ_0=3000 MPa, δ_0=10^{-8} m is plotted again in Fig. D2.6a. This representation gives rise for an approximate method to determine K_{I0} by straight-line fitting over the expanded near crack-tip region.

For this purpose, the deviations of the straight-line, indicating the crack profile in the absence of bridging, and the displacements affected by bridging have to be minimized in an averaged sense. We therefore propose to extrapolate the displacements for

$\delta/\delta_0 > 1.5$ backwards as shown by the dash-dotted line in Fig. D2.6a. The intersection of this line with a straight line drawn from the point $\delta=0$ defines the end of the fitting region. This procedure now yields the two encircled hatched areas (I) and (II). If these areas are identical in size, a least squares fit over homogeneously distributed measurements should result in the correct K_{I0}.

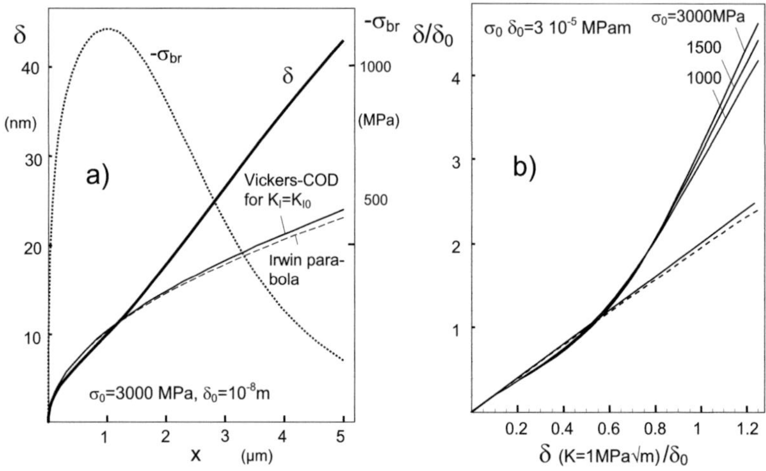

Fig. D2.5 Computations with eq.(D2.4.1): a) crack opening displacements and bridging stress distribution for the parameters σ_0=3000 MPa, δ_0=10⁻⁸ m; b) displacements in presence of bridging plotted versus the displacements for a constant K=1 MPa√m (dashed line: Irwin parabola).

Due to the unavoidable scatter of experimental data it is of course not possible to identify the areas (I) and (II) in practice. Nevertheless, it is recommended to fit the best straight line through the origin and to include all the data points, which are on the left of the intersection point of the fit line and the large displacement extrapolation curve.

This has been done in Fig. D2.6b for the data of Fig. D2.3a. The straight-line fit results in a slope of $\cong 2.1$. Consequently, we obtain $K_{I0} \cong 2.1$ MPa√m with the advantage of very small numerical effort. For small crack-tip distances the COD-field for $K=K_{I0}$ and the approximation by the Irwin parabola differ less than 1.2% in the fitting region (see Figs. D2.5a and D2.5b).

Measurements similar to those on the MgY-Si$_3$N$_4$ were carried out in [D2.11] for the commercial material SL200BG and the hot-pressed Si$_3$N$_4$ SiY (Section C1). In order to study the scatter of K_{I0} and the uncertainties of the displacement measurements, a number of 15 cracks for each material were evaluated resulting in about 1500 (SL200) and 2500 (SiY) individual data points.

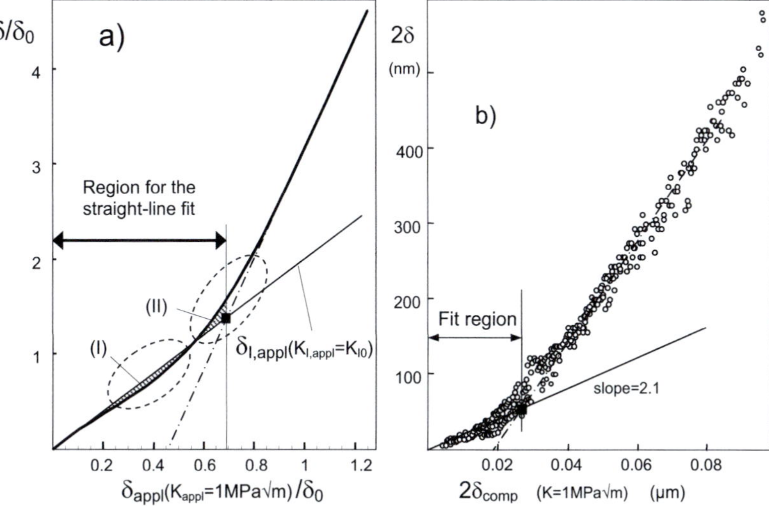

Fig. D2.6 a) Determination of the region for straight-line fitting, b) fit of data from Fig. D2.3.

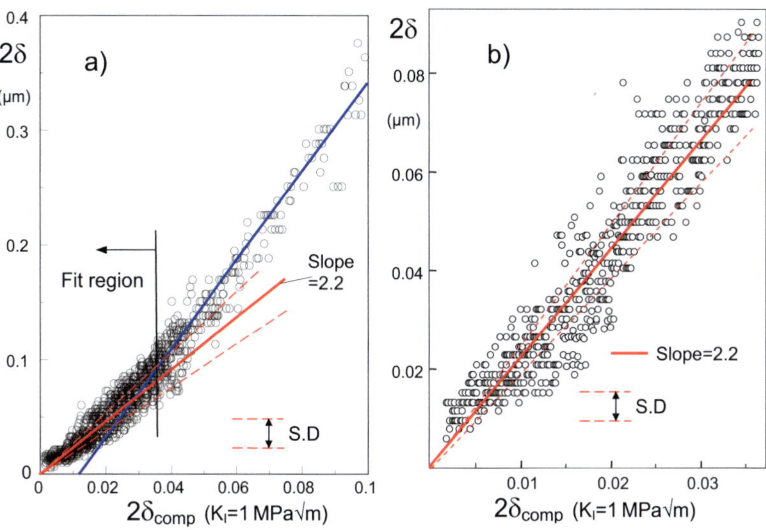

Fig. D2.7 a) Evaluation of 1500 COD-measurements on 15 indentation cracks in the commercial silicon nitride (SL200BG), b) results on the hot-pressed Si_3N_4 SiY.

Figure D2.7a represents the results for SL200. From the least squares fit of the data region indicated by the arrow, a slope of 2.2 was obtained for SL200 yielding K_{I0}=2.2

MPa√m (thick straight line) with a standard deviation of SD=0.66 MPa√m (dashed lines). For SiY the data in the near-tip region are given in Fig. D2.7b. The result is K_{I0}=2.2 MPa√m (straight line) with a standard deviation of SD=0.53 MPa√m.

From all the measurements done by one and the same person and especially from the standard deviations of the slopes we have to conclude a common crack-tip toughness for all the materials of $K_{I0} \approx 2.2 \pm 0.3$ MPa√m.

In this context, it should be mentioned that this value is in good agreement with R-curve results for the SiY-Si_3N_4. In this material with a rather low aspect ratio of about 4 the toughening mechanism by the grains with a high aspect ratio of 7-9 is missing. Its R-curve hardly exceeds the K_{I0}-value as can be seen from Fig. C2.2a. The R-curve values K_R<2.5 MPa√m trivially confirm that K_{I0} must be smaller than 2.5 MPa√m, too.

D2.5 Comparison with COD measurements on straight cracks

Kruzic [D2.12] performed crack opening displacement measurements on the MgY-Si_3N_4. Figure D2.8 shows the measurements carried out with CT-specimens. The solid symbols represent the results for a crack grown under a monotonously increasing load up to K_R=6.76 MPa√m with the measurements carried out at about 97.5% of the maximum load, i.e. at 6.59 MPa√m. In a second test, a cyclic fatigue loading was applied with K_{max}=3.9 MPa√m. After fatigue crack growth at 10^{-10}m/s the specimen was unloaded and COD-measurements performed under 95% of the maximum load at K=3.7 MPa√m. These results are given by the open symbols.

Similar to the procedure described for the Vickers indentation cracks the results of Fig. D2.8 were plotted versus the displacements for a constant stress intensity factor of K=1 MPa√m. For his purpose eq.(B1.4.2) had to be solved with the weight function for the CT-specimen (B1.3.5). In a rewritten form it holds

$$\delta_{appl} = \frac{1}{E'} \int_{a-x}^{a} h_{CT}(x,a')K_{appl,CT}(a')\,da', \qquad (D2.5.1)$$

with $K_{appl,CT}$ given by eq.(B1.3.4).

The result is plotted in Fig. D2.9. For the K_{I0}-determination, Kruzic et al. [D2.13] measured the bridging stress distribution after a fatigue test by Raman spectroscopy. They computed the bridging stress intensity factor with the weight function procedure and subtracted this value from the K_{max} at which the threshold condition of 10^{-10} m/s was reached. The crack-tip toughness after fatigue was obtained as $K_{I0} \approx 1.4$ MPa√m. In order to check this value, a straight line is included in Fig. D2.9 with a slope of 0.95×1.4 =1.33, taking into account that the displacements are measured under the slightly reduced load. Within the data scatter, this line is in agreement with the COD measurements. For the monotonously grown crack, not enough near-tip data were available for an evaluation of K_{I0} under static conditions. Tentatively, the value of $0.975 \times K_{I0} \cong 2.15$ MPa√m was used, resulting in the dashed straight line. This line is at least not in disagreement with the COD data.

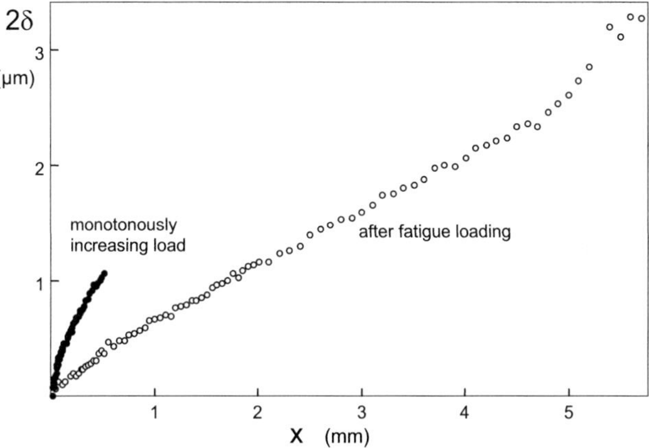

Fig. D2.8 COD for a crack after monotonously increasing load (full circles) and after fatigue loading (open circles) by Kruzic [D2.12].

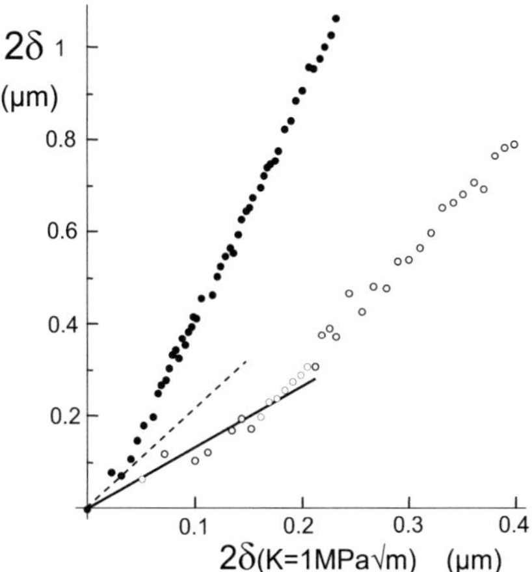

Fig. D2.9 Measured crack opening displacement from Fig. D2.8 plotted versus COD computed for $K=1$ MPa√m.

From the data in Fig. D2.8 also the bridging stresses can be determined. A result is shown in Section C3.4 for the case of the specimen after fatigue.

D2.6 Bridging stresses from Vickers indentations

The evaluation of the crack-tip toughness K_{I0} from the crack opening displacement field of Vickers indentation cracks allows the bridging stress distribution to be determined as a by-product. This may be demonstrated here for the approximation of the Vickers indentation crack by a semi-circular crack with the displacements given by eq. (D2.3.1). In the following computations an average of the weight function for the semi-circular surface crack was used by taking $c=0.42$.

The applied displacements, δ_{appl}, are given by eq.(B2.2.4) and the applied stress intensity factor K_{appl} may be computed from

$$K_{appl} = K_{I0} - \int_0^a h(r,a)\sigma_{br}(r)\,dr \tag{D2.6.1}$$

with the weight function h according to eq.(B2.1.3) and (B2.1.4).

The bridging stresses were determined from the data of Fig. D2.10a by solving the integral equation system, (D2.3.1) and (D2.6.1) and matching the computed to the measured displacements. The computed displacements resulting from the solution of the integral equations are entered in Fig. D2.10a as the curve.

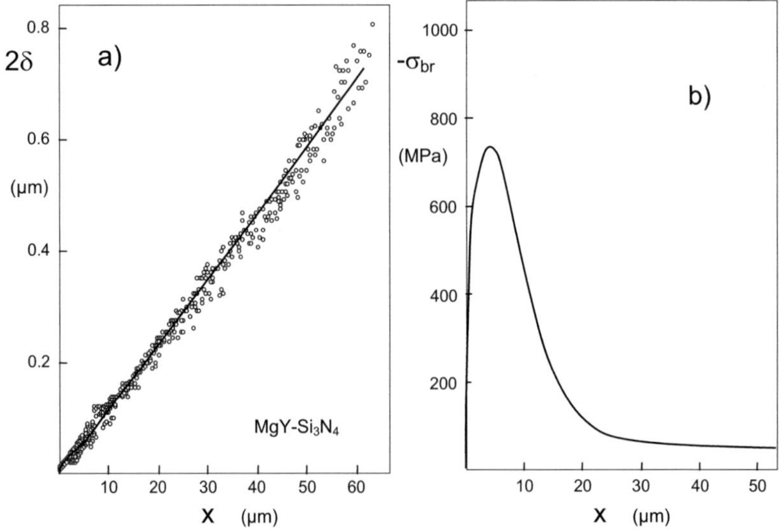

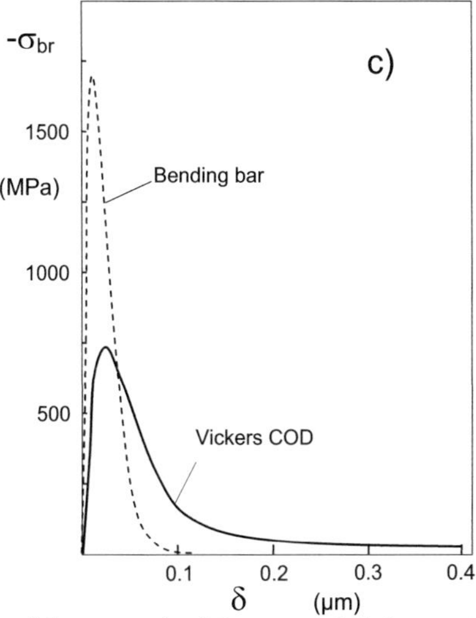

Fig. D2.10 a) Comparison of the measured and the computed displacements, b) bridging stresses as a function of crack-tip distance, c) comparison of the bridging relations obtained from the R-curve (dashed line) and the COD-evaluation (solid line).

The related bridging stresses are shown in Fig. D2.10b as a function of the distance from the tip. Combining Figs. D2.10a and D2.10b yields the bridging law as plotted in Fig. D2.10c. The local shielding stress intensity factor at the surface was found as $K_{sh}=$ −3.5 MPa√m and, consequently, the local value of K_R as $K_R=K_{I0}-K_{sh}=5.7$ MPa√m.

Whereas the bridging stresses from R-curve measurements (dashed curve) reach maximum value of about 1700MPa, from the COD-evaluation (solid curve) a clearly lower peak value of about 750 MPa was obtained. Nevertheless, the areas under the σ_{br}-δ-curves are comparable and the K_R-values are nearly the same. For a possible explanation of this result, see Section C3.4.

References D2

D2.1 J. Rödel, J.F. Kelly, and B.R. Lawn, "In situ measurements of bridged crack interfaces in the scanning electron microscope", *J. Am. Ceram. Soc.* **73**(1990), 3313-18.

D2.2 J. Seidel and J. Rödel, "Measurement of crack tip toughness in alumina as a function of grain size", *J. Am. Ceram. Soc.* **80**(1997), 433-438.

D2.3 T. Fett, D. Munz, J. Seidel, M. Stech, and J. Rödel, "Correlation between long and short crack R-curves in alumina using the crack opening displacement and fracture mechanical weight function approach, *J. Am. Ceram. Soc.* **79**(1996), 1189-96.

D2.4 Kruzic, J.J., Ritchie, R.O., Determining the toughness of ceramics from Vickers indentations using the crack-opening displacements: An experimental study, J. Am. Ceram. Soc. **86**(2003), 1433-36.

D2.5 Pezzotti, G., Muraki, N., Maeda, N., Satou, K., Nishida, T., In situ measurement of bridging stresses in toughened silicon nitride using Raman microprobe spectroscopy, J. Am. Ceram. Soc. **82**(1999), 1249-56

D2.6 F. Meschke, O. Raddatz, A. Kolleck, and G.A. Schneider, "R-curve behaviour and crack-closure stresses in barium titanate and (Mg,Y)-PSZ ceramics", *J. Am. Ceram. Soc.* **83**(2000), 353-61.

D2.7 Meschke, F., Alves-Ricardo, P., Schneider, G.A., Claussen, N., Failure behaviour of alimina and alumina/silicon carbide nanocomposite with natural and artificial flaws, J. Mater. Res. **12**(1997), 3307-15.

D2.8 Burghard, Z., Zimmermann, A., Rödel, J., Aldinger, F., Lawn, B.R., Crack opening profiles of indentation cracks in normal and anomalous glasses, Acta Mater. **52/2**(2004), 293-97.

D2.9 Kounga Njiwa, A.B., Fett, T., Rödel, J., Quinn, G.D., Crack-tip toughness measurements on a sintered reaction-bonded Si3N4, J. Am. Ceram. Soc. **87**(2004), 1502-1508.

D2.10 Fett, T., Kounga Njiwa, A.B., Rödel, J., Crack opening displacements of Vickers indentation cracks, Engng. Fract. Mech. **72**(2005), 647-659.

D2.11 S. Fünfschilling, T. Fett, R. Oberacker, M.J. Hoffmann, G.A. Schneider, P.F. Becher, J.J. Kruzic, Crack-tip toughness from Vickers crack-tip opening displacements for materials with strongly rising R-curves, J. Am. Cer. Soc., **94**(2011), 1884–1892.

D2.12 Kruzic, J.J., unpublished results.

D2.13 Kruzic, J.J., Cannon, R.M., Ager III, J.W., Ritchie, R.O., Fatigue threshold R-curves for predicting reliability of ceramics under cyclic loading, Acta Mater. **53**(2005), 2595-2605.

E1 Residual stress intensity for indentation cracks

E1.1 Residual stress intensity factor after load removal

Knoop indentation tests carried out on brittle materials are accompanied by the generation of a half-elliptical surface crack below the indenter, which fully develops during
unloading by the action of residual stresses. During loading by a Knoop indenter a residual stress zone develops below the contact area. According to the model proposed
by Marshall [E1.1] a prolate spheroid can be chosen for the shape of the irreversibly
deformed 'plastic' zone with the major axis b_1 and the depth b_2 (Fig. E1.1a).

If σ_{res} is the residual stress assumed to be constant over the semi-elliptic cross section
with the half-axes b_1 and b_2, the total force normal to the crack plane is

$$P_{res} = \tfrac{1}{2}\sigma_{res}\pi\, b_1 b_2 \qquad\qquad (E1.1.1)$$

For the fracture mechanics analysis of Knoop indentation cracks, the stress intensity
factor caused by the residual stress field is necessary. Keer et al. [E1.2] computed
stress intensity factors for a few special geometries. If the residual stress intensity factors at the deepest point (A) and the surface points (B) are denoted as $K_{A,B}$, it holds

$$K_{A,B} = \frac{2P_{res}}{(\pi a)^{3/2}} F_{A,B} \qquad\qquad (E1.1.2)$$

with the geometric functions F_A and F_B given by eqs.(I1.3.3) and (I1.3.4).
So far residual forces P_{res} are considered acting normal to the crack plane. In a Knoop
indentation tests, the load P acting *in* indentation direction is given. Using the proportionality between P_{res} and P the stress intensity factor can be written

$$K_{A,B} = \chi\left(\frac{E}{H}\right)^{1/2}\frac{P}{(\pi a)^{3/2}} F_{A,B} = \chi\left(\frac{E}{H}\right)^{1/2}\frac{P}{(\pi c)^{3/2}}\left(\frac{c}{a}\right)^{3/2} F_{A,B} \qquad (E1.1.3)$$

adopting the ratio E/H according to Lawn et al.[E1.3] with an a priori unknown parameter χ.

E1.2 Evaluation of experimental results

Many experimental investigations on Knoop indentation cracks are known from literature. Only in a few cases all the parameters b_1, b_2, a/c and a (or c) are reported that are
necessary for the computation of the stress intensity factor. In a paper by Lube [E1.4]

the silicon nitride EkasinS (ESK, Kempten) was investigated from which all dimensions are available necessary for an evaluation of eq.(E1.1.3). Knoop indentations were made for a load range of 29 N$\leq P \leq$294 N and the dimensions of decorated cracks measured on the crack face as well as by sectioning from the side normal to the crack plane. The resulting data are plotted in Fig. E1.1b. The axis ratio of the residual stress zone was found to be $b_1/b_2 \cong 3 \pm 0.5$. The length of the deformed zone was found to be somewhat smaller than the large indentation diagonal, namely, $2b_1 \cong 0.8L$. For the computation of the ordinate values in Fig. E1.1c, the experimental data for b_1 were used together with $b_1/b_2 = 3$.

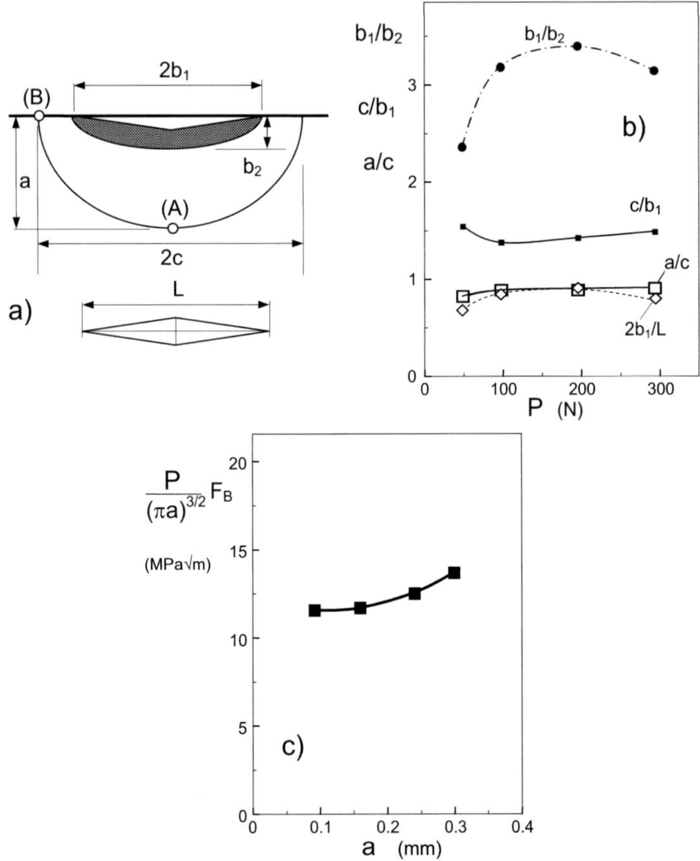

Fig. E1.1 a) Semi-elliptical crack loaded by a residual stress zone of length $2b_1$ and width b_2 as suggested by Marshall [E1.1]; b) dimensions of crack and deformation zone for Knoop cracks [E1.4] in MgO-containing Si_3N_4 (Ekasin-S) versus indentation load; c) results for the surface points.

Since b_1 and b_2 were measured directly, an assumption of $b_1=L/2$ was not necessary for the stress intensity factor evaluation. The results on further silicon nitrides are shown in Fig. E1.2. A slight increase of the stress intensity factors with increasing crack length a is visible.

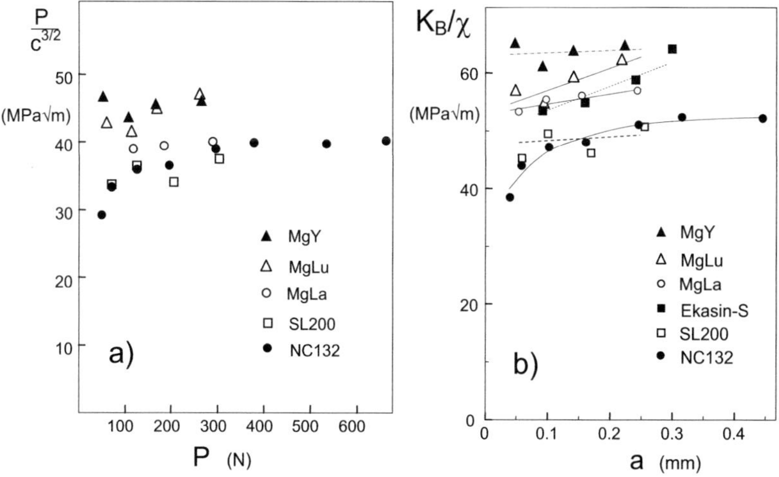

Fig. E1.2 Measurements on several silicon nitrides, a) $P/c^{3/2}$ vs. load P, b) normalized stress intensity factor according to eq.(E1.1.3).

The solid circles represent early experimental results reported by Marshall [E1.1]. The other materials (see Section C1) were investigated by Fünfschilling [E1.5]. Figure E1.2a shows the original data in a plot of $P/c^{3/2}$ vs. P. Since the aspect ratios a/c were not measured for these materials, averages of the results by Lube [E1.4] were applied. As tenable parameters the average values of $a/c=0.85$ and $b_1=0.8\,L/2$ were used. Figure E1.2b gives the results in form of eq.(E1.1.3) as a function of the crack depth with the data for Ekasin-S by Lube [E1.4] included.

Relating the stress intensity factor at point (B) to the actual crack-growth resistance at the same location, $K_R(c)$, establishes

$$K_{(B)} = K_R(c) \quad \Rightarrow \quad \chi\left(\frac{E}{H}\right)^{1/2}\frac{P}{(\pi a)^{3/2}}F_B = K_R(c) \qquad (E1.2.1)$$

The unknown parameter χ may be determined from a material without an R-curve effect. In such a case, the parameter should be independent on the crack length. A possibility would be glass not even characteristic for high-toughness ceramics. A more feasible calibration would be the use of a material with strongly rising R-curve, reaching saturation resistance even after a few micrometers. Such an R-curve exists for the Y-Lu-silicon nitride and the SL200 (see Fig. C2.2). The results of Fig. E1.2b may be used for the parameter χ to be estimated from a least squares fit. Identifying the width c of the Knoop crack with the crack extension Δa of the R-curve yields the related crack resistance $K_B=K_R$ for the surface point of the indentation crack. Figure E1.3 shows the dependency $K_R=f(K_B/\chi)$ with K_R taken from Fig. C2.2. From the slope of the straight-line, it results

$$\chi \cong 0.11 \pm 10\% \qquad\qquad (E1.2.2)$$

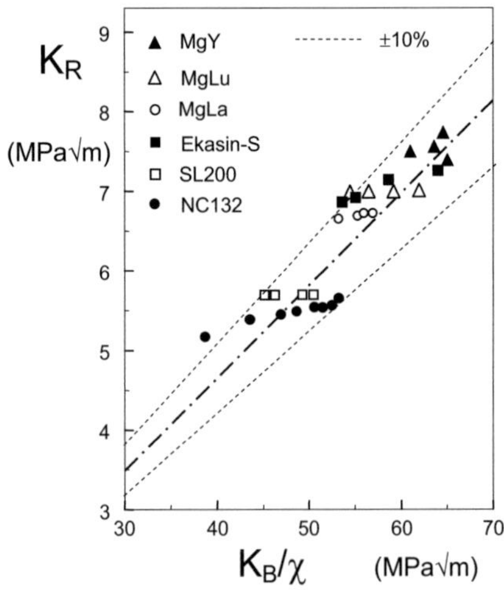

Fig. E1.3 Crack resistance K_R for $\Delta a = c$; dash-dotted line through the origin represents the coefficient χ of (E1.2.2), dashed lines indicate the $\pm 10\%$ scatter band.

References E1

E1.1 Marshall, D.B., Controlled flaws in ceramics: A comparison of Knoop and Vickers indentation, J. Am. Ceram. Soc. **66**(1983), 127-131.

E1.2 Keer, L.M., Farris, T.N., Lee, J.C., Knoop and Vickers indentation in ceramics analyzed as a three-dimensional fracture, J. Am. Ceram. Soc. **69**(1986), 392-96.

E1.3 Lawn, B.R., Evans, A.G., Marshall, D.B., Elastic/plastic indentation damage in ceramics: The medial/radial crack system, J. Am. Ceram. Soc. **63**(1980), 574-81.

E1.4 Lube, T., Indentation crack profiles in silicon nitride, J. Europ. Ceram. Soc. **21**(2001), 211-218.

E1.5 Fünfschilling, S., Influence of microstructure on the R-curve behaviour of silicon nitride ceramics (in German), Thesis, IKM 53, KIT Scientific Publishing, Karlsruhe, 2009; (Open Access: http: /creativecommons.org.lizenses/by-nc-nd/3.0/de/).

E2 Partial unloading of the residual stress term

E2.1 Superposition of residual and applied stress intensity factors

In an indentation test on brittle material a system of cruciform cracks is generated. The development of the half-elliptical surface crack below the indenter is fully finished after unloading by the action of residual stresses which are responsible for a residual stress intensity factor K_{res} (see especially Section D2 for a Vickers indentation test). In the absence of subcritical crack growth it holds after the indentation test at the deepest point (A) and the surface points (B)

$$K_{res,A,B} = K_{Ic} \qquad (E2.1.1)$$

If such a crack is additionally loaded by an external tensile stress σ normal to the crack plane, an applied stress intensity factor is generated which is given by

$$K_{appl,A,B} = \sigma_{appl} F_{appl,A,B}(a/c)\sqrt{\pi a} \qquad (E2.1.2)$$

with the crack depth a and the geometric function $F_{appl,A,B}$ depending on the aspect ratio a/c (see Section I1).
The total stress intensity factor $K_{A,B}$ is often computed by superimposing the two stress intensity factor terms

$$K_{A,B} = K_{res,A,B} + K_{appl,A,B} \qquad (E2.1.3)$$

Such an attempt of course supposes that no interaction between the applied and the residual terms will occur.
A decreasing residual stress intensity factor under an external load can be concluded from experiments (see e.g. Section E3) and theoretical considerations. In the simplest limit case of purely displacement-controlled residual crack opening (i.e. a rigid wedge opens the crack), a decreasing residual stress intensity factor with an increasing applied load could be predicted [E2.1] establishing a reduction factor of

$$\frac{K_{res}}{K_{res,0}} = \kappa \neq 0 \qquad (E2.1.4)$$

with the residual stress intensity factor in the absence of a superimposed applied load, $K_{res,0}$.

In order to compute the dependency $K_{res}(K_{appl})$ for the more realistic case of a constant strain in the damage zone below the indenter, a finite element analysis was performed.

E2.2 Finite Element computations

The case of a penny-shaped crack in the infinite solid was realized by a finite element mesh of about 2000 elements with 6300 nodes and a component height of $H=100a$ and component width of $W=100a$. The computations were carried out with ABAQUS Version 6.2.

Two different models of the load were studied. First, a constant volume strain in the damaged zone was investigated. In order to apply the option for computing thermal stress intensity factors with ABAQUS, the volumetric strain in the damaged zone below the Vickers indenter can be replaced by a thermal strain and the stresses by thermal stresses due to different temperature of inclusion and matrix

$$\varepsilon_{vol} = \frac{\Delta V}{V} \equiv 3\alpha\Delta T \tag{E2.2.1}$$

as illustrated in the picture above Fig. E2.1a.

Figure E2.1a shows the FE-result for the circular crack as the reduction factor κ of the residual stress intensity factor according to (E2.1.4) as a function of the applied stress intensity factor. From this representation, it becomes obvious that the residual stress intensity factor for constant geometry decreases for increased applied stress intensity factors.

In a second attempt, only the tractions normal to the crack plane were considered. The results are shown in Fig. E2.1b. It may be astonishing that the two attempts do not show identical results. Whereas the prescribed tractions and the residual stress intensity factor disappear when the crack faces are completely separated, the prescribed strains in Fig. E2.1a result in negative stress intensity factors. The reason is the occurrence of a stress component parallel to the crack plane. It generates a negative K-contribution as directly follows from the analysis of McMeeking and Evans [E2.2].

The FE-data of Fig. E2.1b (circles) were fitted by

$$\kappa = \frac{K_{res}}{K_{res,0}} = 1 - \tanh^{1/2}\left[0.22(1+\tfrac{1}{2}\frac{a}{b})\left(\frac{b}{a}\right)^3\frac{K_{appl}^2}{K_{res,0}^2}\right] \tag{E2.2.2}$$

where $K_{res,0}$ denotes the residual stress intensity factor in the absence of K_{appl}. Equation (E2.2.2) is introduced in Fig. E2.1b by the curves.

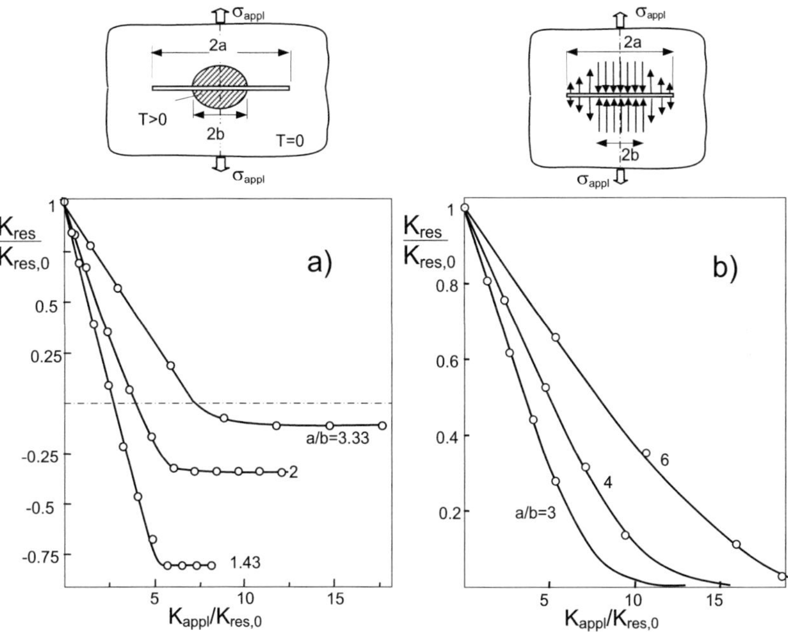

Fig. E2.1 a) Circular crack in an infinite body loaded by an extending sphere with volumetric strain ε_{vol} and superimposed load σ_{appl}, b) residual stress intensity factor K_{res} for loading by normal tractions exclusively.

E2.3 Conditions for stable crack growth

Predictions for the stable crack-extension phase can be made for the Vickers indentation cracks using the total stress intensity factors K_A and K_B including the effect of partial reduction of the residual stress intensity factor term under a superimposed applied stress intensity factor by solving the equation system

$$K_{appl,A} + K_{res,A} = \sigma_{appl}F_{appl,A}\sqrt{\pi a} + \chi \frac{P}{a^{3/2}}F_{res,A}\kappa_A = K_{Ic} \tag{E2.3.1}$$

$$K_{appl,B} + K_{res,B} = \sigma_{appl}F_{appl,B}\sqrt{\pi a} + \chi \frac{P}{a^{3/2}}F_{res,B}\kappa_B = K_{Ic} \tag{E2.3.2}$$

with the geometric functions F_{res} for loading by the residual stress field and

$$\kappa_{A,B} \cong 1 - \tanh^{1/2}\left[0.22(1 + \tfrac{1}{2}\frac{a}{b})b^3 \frac{K^2_{appl,A,B}}{(\chi P)^2} \right]$$

(E2.3.3)

References E2

E2.1 Fett, T., Fracture of ceramics with surface flaws introduced by Knoop indentation, J. Mat. Sci. **19**(1984),672-682.
E2.2 McMeeking, R.M., Evans, A.G. Mechanics of transformation toughening in brittle materials, J. Am. Ceram. Soc. **65**(1982), 242–246.

E3 Residual stress intensity during crack extension

Beneath the contact area of an indenter pressed into a ceramic surface, a residual stress zone remains even after unloading. Figure E3.1a gives the relevant geometric data of the cruciform Vickers crack system generally modelled by a single semi-elliptical surface crack. The parameter b is the radius of the damaged zone responsible for the residual stresses assumed semi-circular for Vickers and semi-elliptical for Knoop indentations (Fig. E3.1b) with half axes b_1 and b_2. These residual stresses are responsible for a residual stress intensity factor K_{res}. It is the aim of this section to determine the stress intensity factor K_{res} during the stable crack extension phase under a superimposed bending load.

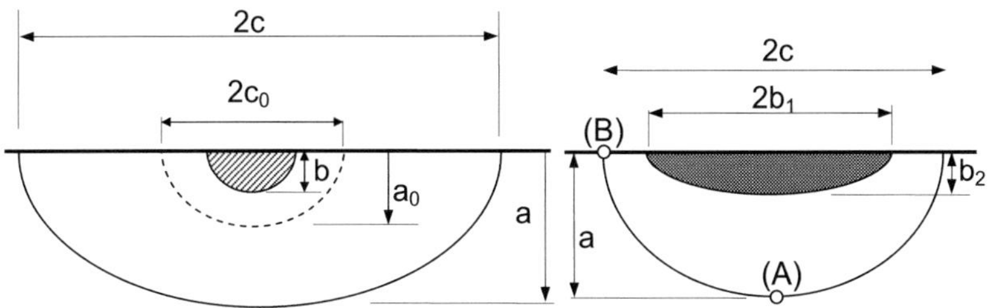

Fig. E3.1 Geometric data of indentation cracks: a) Vickers indentation crack of initial width $2c_0$ grown under an externally applied bending load to the width $2c$, b) Knoop indentation crack.

Extensive results on stable crack extension of silicon nitrides are available in literature. In the following, the data by Marshall [E3.1] and Lube [E3.2, E3.3] will be addressed. Marshall considered the NC132 (Norton, Seven Oaks, USA) and Lube the gas pressure sintered silicon nitride Ekasin-S (ESK, Kempten, Germany).

E3.1 Vickers-indentation cracks

Figure Fig. E3.2a shows measurements of the crack aspect ratio a/c as a function of crack width c normalized on the thickness t of the bending bars. The results on Ekasin-S, measured on decorated cracks, are represented for two indentation loads of P=49N and 196N by the solid symbols. These indentation cracks were impregnated with lead-acetate, loaded after drying and extended. At an increased load the extended cracks were impregnated again. After drying the load was increased up to fracture strength. So, the initial crack dimensions, those during crack propagation, and finally, the critical size could be determined. Results of final dimensions in NC132 are entered as squares.

The dashed curve shows the equilibrium aspect ratio, i.e. that aspect ratio for which the stress intensity factor at the surface and the deepest point of the semi-ellipse are identical [E3.4]

$$\left(\frac{a}{c}\right)_{eq} = \frac{1}{(c/t)^2}\left[3.72 + 5.141\frac{c}{t} - 4.516\sqrt{0.6787 + 1.876\frac{c}{t} + \left(\frac{c}{t}\right)^2}\right] \qquad \text{(E3.1.1)}$$

This aspect ratio establishes an asymptotic lower limit solution. For any initial aspect ratio a_0/c_0, the ratio a/c will tend to this line.

A rough fit relation of the strongly scattering data is

$$\frac{a}{c} \cong \left(\frac{a}{c}\right)_{eq} + \left(\left(\frac{a_0}{c_0}\right) - \left(\frac{a}{c}\right)_{eq}\right)\exp[-\lambda(c - c_0)/t] \qquad \text{(E3.1.2)}$$

with $\lambda=40$, plotted for $P=49N$ and 196N by the solid curves. The same relation was used for the NC132 results, tentatively introduced by the dash-dotted curve.

Under a superimposed externally applied load (in our case bending) the applied stress intensity factor at the surface is given by

$$K_{appl,B} = \sigma_{appl}F_{appl,B}\sqrt{\pi a} \qquad \text{(E3.1.3)}$$

with the geometric function $F_{\text{appl, B}}$. The geometric function reads for $0.7 \leq a/c \leq 1$ and $a/t \leq 0.15$ [E3.4, E3.5]

$$F_{appl,B} = 0.755 + 0.0232\frac{a}{c} - 0.0714\left(\frac{a}{c}\right)^2 - 0.569\frac{a}{t} + 0.739\left(\frac{a}{t}\right)^2$$

$$+ 0.0528\frac{a}{c}\frac{a}{t} - 0.541\left(\frac{a}{c}\right)\left(\frac{a}{t}\right)^2 \qquad \text{(E3.1.4)}$$

computed according to the weight function procedure by Cruse and Besuner [E3.6]. Figure E3.2b gives the crack extension data in the form of $K_{\text{appl}}=f((c_0/c)^{3/2}$. The shaded areas indicate the crack resistance for the interesting range of crack growth of about 70-500μm.

Stable crack growth is governed by the total stress intensity factor, which results by superposition of the applied stress intensity factor K_{appl} and the residual stress intensity factor term K_{res}. At the surface (point B) this requires

$$K_{total,B} = K_{res,B} + K_{appl,B} = K_R(c) \qquad \text{(E3.1.5)}$$

Under the assumption that K_{res} and K_{appl} do not interact, the residual K-value would be of the type

$$K_{res} = K_{Ic}\left(\frac{c_0}{c}\right)^{3/2}$$

(E3.1.6)

with the initial crack length c_0 and K_{Ic} as a representative average of the K_R-data (5.0±0.1 MPa√m for NC132 and 6.5±0.1 MPa√m for Ekasin). Consequently, a plot of K_{appl} vs. $(c_0/c)^{3/2}$ should result in a straight line as for instance plotted in Fig. E3.2b for NC132 (solid line). The measured results (circles) clearly deviate from this line. The same holds for the Ekasin (dashed straight line).

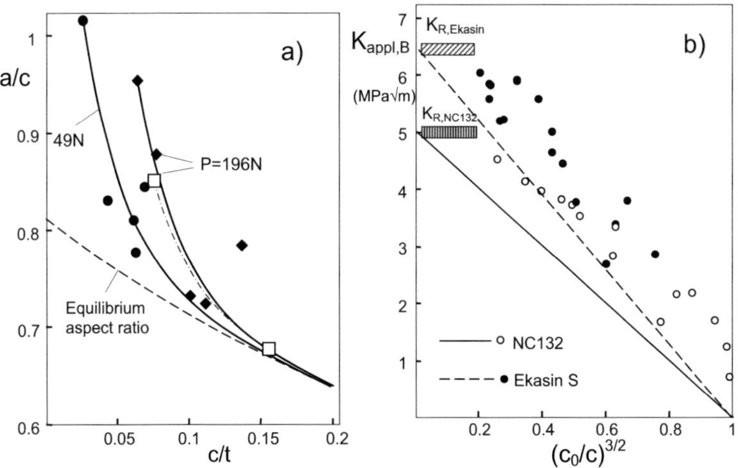

Fig. E3.2 a) Development of aspect ratios a/c during Vickers-crack extension; solid symbols: Ekasin-S, squares: Norton NC132; b) experimental crack growth results for Vickers-cracks in Si_3N_4 with applied stress intensity factors computed from eqs.(E3.1.3) and (E3.1.4); straight lines: predictions for a residual stress term according to eq.(E3.1.6).

E3.2 Knoop-indentation cracks

E3.2.1 Experimental results

Figure E3.3 gives the measured data for the Knoop indentation cracks. The aspect ratios as a function of relative crack width c/t are plotted in Fig. E3.3a. These curves tend rather slowly to the bending aspect ratio resulting in $\lambda=18$ for eq.(E3.1.2).

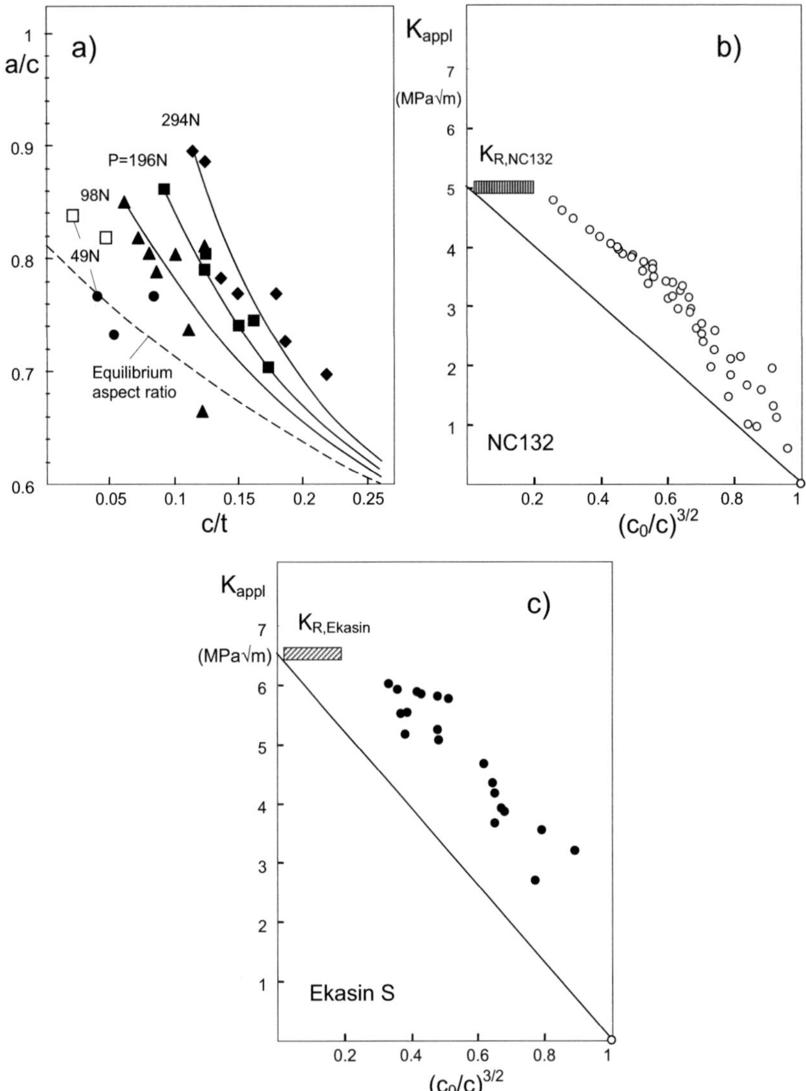

Fig. E3.3 a) Aspect ratios *a/c* during Knoop-crack extension; b) experimental crack growth results for Knoop-cracks in NC132 computed with eqs.(E3.1.3) and (E3.1.4), c) results for Ekasin-S; straight lines: predictions for a residual stress term according to eqs.(E3.1.5, E3.1.6).

Figure E3.3b again gives the crack extension data in the form of $K_{appl}=f(c_0/c)^{3/2}$ for NC132 and E3.3c for Ekasin-S. Also for the Knoop indentation cracks, the deviations from the predicted straight lines are clearly visible.

E3.2.2 Computation of the residual stress intensity factor

Since $K_R(c)$ and $K_{appl,B}$ are known, eq.(E3.1.5) allows the residual stress intensity factor $K_{res,B}$ to be determined as a function of crack length.

The crack resistance $K_R(c)$ for indentation cracks is represented in Fig. E3.4 for the two ceramics. These data were derived from results obtained with bending bars (Fig. C2.2) by application of the weight function procedure according to [E3.7, E3.8].

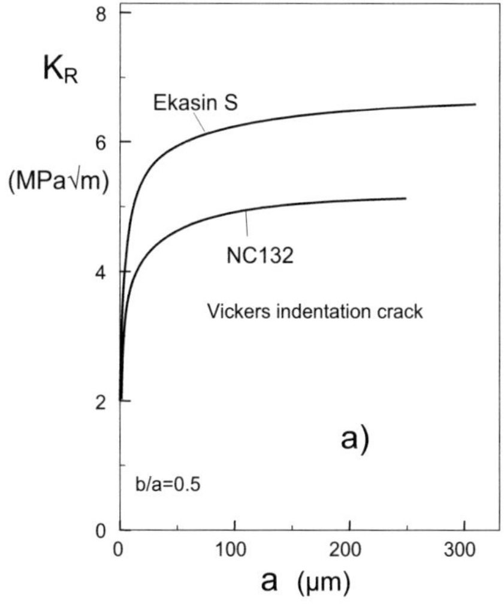

Fig. E3.4 Crack-resistance curves for indentation cracks computed from results on deep-notched bending bars by using the weight function procedure according to [E3.7, E3.8] (a=crack length, b=radius of the residual stress zone).

Figure E3.5a shows the residual stress intensity term normalized on crack-growth resistance $K_R(c)$ as a function of $(c_0/c)^{3/2}$. The data for both materials are clearly below the dash-dotted straight-line that would reflect the proportionality (E3.1.6). Figure E3.5b shows the same plot for the Knoop cracks.

Comparison of Figs. E3.5a and E3.5b shows the same behaviour, namely a clear decrease of the residual stress intensity factor for both types of cracks [E3.9].

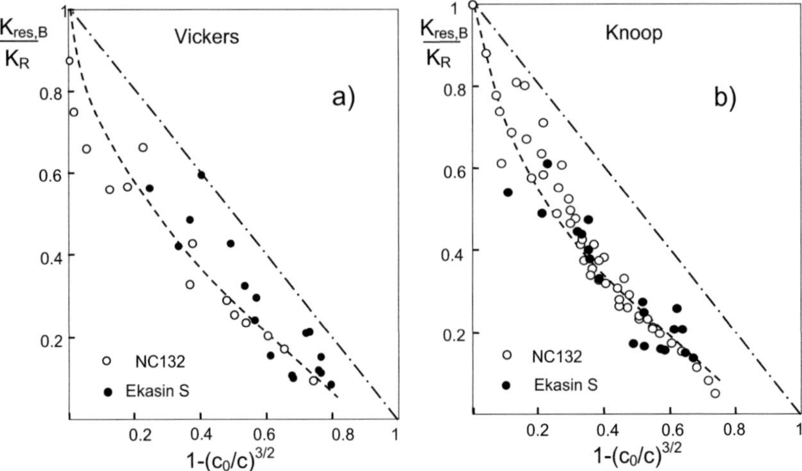

Fig. E3.5 Residual stress intensity factor at the surface of indentation cracks as a function of crack width ratio c/c_0, a) for Vickers cracks, b) for Knoop cracks; dash-dotted lines: predicted from eqs.(E3.1.5, E3.1.6).

References E3

E3.1 Marshall, D.B., Controlled flaws in ceramics: A comparison of Knoop and Vickers indentation, J. Am. Ceram. Soc. **66**(1983), 127-131.

E3.2 Lube, T., Indentation crack profiles in silicon nitride, JECS **21**(2001), 211-218.

E3.3 Lube, T., Investigation on the stable crack growth of indentation cracks, in: J.G. Heinrich, F. Aldinger (Eds.), Ceramic Materials and Components for Engines, Wiley-VCH, Weinheim, 2001, 121-126.

E3.4 Fett T. Stress intensity factors, T-stresses, Weight function (Supplement Volume), IKM55, KIT Scientific Publishing, Karlsruhe; 2009.

E3.5 T. Lube, T. Fett, S. Fünfschilling, M. Hoffmann, R. Oberacker, A residual stress intensity factor solution for Knoop indentation cracks, In. J. Fract (in press).

E3.6 Cruse, T.A., Besuner, P.M., Residual life prediction for surface cracks in complex structural details, J. of Aircraft **12**(1975), 369-375.

E3.7 Fett, T., Fünfschilling, S., Hoffmann, M. J., Oberacker, R., Different R-curves for 2- and 3-dimensional cracks, Int. J. Fract. **153**(2008), 153-159.

E3.8 Fünfschilling, S., Fett, T., Hoffmann, M. J., Oberacker, R., Schwind, T., Wippler, J., Böhlke, T., Özcoban, H., Schneider, G.A., Becher, P.F., Kruzic, J.J., Mechanisms of toughening in silicon nitrides: The roles of crack bridging and microstructure, Acta Materialia **59**(2011), 3978-89.

E3.9 Lube, T., Fett, T., A threshold stress intensity factor at the onset of stable extension of Knoop indentation cracks, Engng. Fract. Mech. **71**(2004), 2263-2269.

F1 Influence of the T-stress on frictional bridges

The first regular term of the crack-tip stress field gives rise for two effects on the R-curve behaviour. The first effect, a reduced debonding has been discussed in Section C3.4. The second influence on the frictional bridges may be discussed in this section.

The bridging effects by frictional or elastic crack-face interlocking were described in Section C3.3 for an isotropic state of intrinsic mismatch stresses by application of the bridging model by Mai and Lawn [F1.1]. In a more general treatment, the different mismatch stress components may be considered.

In the schematic depiction of Fig. F1.1, a large grain is shown, acting as a frictional crack-bridging event. The x-component of the thermal mismatch tractions σ_{mis} is indicated. For a 3-dimensional analysis, we have to consider crack-face interlocking with a finite depth L in the order of $L \approx D$. The consequence is that mismatch stresses also act in y-direction of Fig. F1.1 (y-component not plotted in Fig. F1.1b).

The loads transferred by crack face interactions are localized at single grains. The bridging stress σ_{br} for such an element can be expressed by

$$\sigma_{br} \cong \begin{cases} \mu(\sigma_{mis,x} + \sigma_{mis,y}) & for & \sigma_{mis,x} < 0,\ \sigma_{mis,y} < 0 \\ \mu\sigma_{mis,x} & for & \sigma_{mis,x} < 0,\ \sigma_{mis,y} > 0 \\ \mu\sigma_{mis,y} & for & \sigma_{mis,x} > 0,\ \sigma_{mis,y} < 0 \\ 0 & & else \end{cases} \qquad (\text{F1. 1})$$

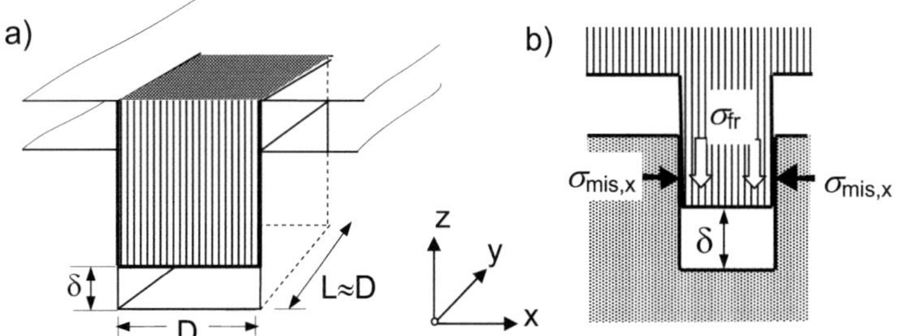

Fig. F1.1 Crack surface interactions due to a local frictional bridging event, a) geometric data, b) tractions, acting at an interlocking event (tractions in y-direction not plotted).

As shown in Section C3.3, the "global" tractions for frictional bridging are

$$\sigma_{br} = \sigma_0 \exp(-\delta / \delta_0) \qquad (\text{F1.2})$$

with the two a priori unknown material specific parameters σ_0 and δ_0.

The x-stress component in a cracked body is

$$\sigma_x = \frac{K_1}{\sqrt{2\pi r}}\, f(\varphi) + T + O(r^{1/2}) \tag{F1.3}$$

with the origin of the polar coordinates $(r,\ \varphi)$ located at the crack tip. The angular function $f(\varphi)$ is maximum ahead of the tip ($\varphi=0$) and disappears in the crack wake ($\varphi=\pi$). Therefore, the near-tip stresses at a bridging event in the crack wake are

$$\sigma_x = \sigma_{mis,x} + T \tag{F1.4}$$

$$\sigma_y = \sigma_{mis,y} \tag{F1.5}$$

For edge-cracked specimens with the crack depth a much smaller than the specimen thickness W under a remote stress σ_n normal on the crack faces the T-stress is negative (compression) and it holds simply

$$T = -\lambda\,\sigma_n \tag{F1.6}$$

with $\lambda=0.5259$ (see e.g. [F1.2]). Depending on the relative crack depth $\alpha = a/W$ the T-stress can be tension and compression for one and the same external load. As a consequence of eq.(F1.4), we see that the T-stress, caused by the applied load, can lower or strengthen the effect of thermal mismatch.

In terms of the bridging stress relation, eq.(F1.1), the characteristic bridging stress σ_0 then holds

$$\sigma_0 = \sigma_{00}(1 + c\lambda\sigma_n) \tag{F1.7}$$

where $c>0$ is an a priori unknown material specific parameter and σ_{00} stands for the characteristic bridging stress in the absence of an additional stress, i.e. σ_{00} represents the thermal mismatch exclusively. Consequently, the bridging stress intensity factor reads

$$K_{br} = K_{br,0}(1 + c\lambda\sigma_n) \tag{F1.8}$$

($K_{br,0}=K_{br}$ for $T=0$). It increases for $T<0$ and decreases for $T>0$ [F1.3].

References F1

[F1.1] Mai, Y., Lawn, B.R., Crack-interface grain bridging as a fracture resistance mechanism in ceramics: II. Theoretical fracture mechanics model, J. Am. Ceram. Soc. **70**(1987), 289.

[F1.2] Fett, T., Stress intensity factors, T-stresses, weight functions, Universitätsverlag Karlsruhe (open access under: http://creativecommons.org/licenses/by-nc-nd/2.0/de/), 2008, Karlsruhe.

[F1.3] Fett, T., Friction-induced bridging effects caused by the T-stress, Engng. Fract. Mech. **59** (1998), 599-606.

F2 A possible influence of multi-axial stresses

F2.1 Frictional bridging under external multi-axial stresses

In section F1, we considered only cracks, which are loaded by remote tractions exclusively normal to the crack plane. In addition to the T-stress let us now consider the existence of externally applied mechanical stresses which can act in x and y-direction. These *externally applied* tractions may be denoted here as $\sigma_{ext,x}$ and $\sigma_{ext,y}$, respectively. The effect of this external load has to be added to the effects of thermal mismatch stress and T-stress.

The characteristic bridging stress σ_0 in eqs.(F1.2) and (F1.7) then reads

$$\sigma_0 = \sigma_{00}(1 + c[\lambda\sigma_n - \sigma_{ext,x} - \sigma_{ext,y}]) \qquad (F2.1.1)$$

again with $c>0$ and the coefficient λ defined by eq.(F1.6). Finally, the bridging stress intensity factor results as

$$K_{br} = K_{br,0}(1 + c[\lambda\sigma_n - \sigma_{ext,x} - \sigma_{ext,y}]) \qquad (F2.1.2)$$

F2.2 Biaxial tension loading

Biaxial tension with the principal stresses $\sigma_1=\sigma_2$ and $\sigma_3=0$ can be realized in a ring-on-ring test [F2.1]. In this test it holds $\sigma_{ext,y}=\sigma_{ext,z}=\sigma_1$ and $\sigma_{ext,x}=\sigma_3=0$. A crack at the tensile surface is assumed as illustrated in Fig. F2.1a. The z-stress $\sigma_{ext,z}$ might be the normal stress σ_n on the crack plane (Fig. F2.1b). A bridging interaction is indicated by the hatched cuboid. In this case the stress component $\sigma_{ext,y}$ is responsible for an additional bridging effect.

$$\sigma_0 = \sigma_{00}[1 + c(\lambda\sigma_n - \sigma_{ext,y})] = \sigma_{00}[1 + c\sigma_1(\lambda - 1)] \qquad (F2.2.1)$$

with the result of

$$K_{br} = K_{br,0}[1 + c\sigma_1(\lambda - 1)] \qquad (F2.2.2)$$

Since in this test $\sigma_1 > 0$, the bridging effect is lowered.

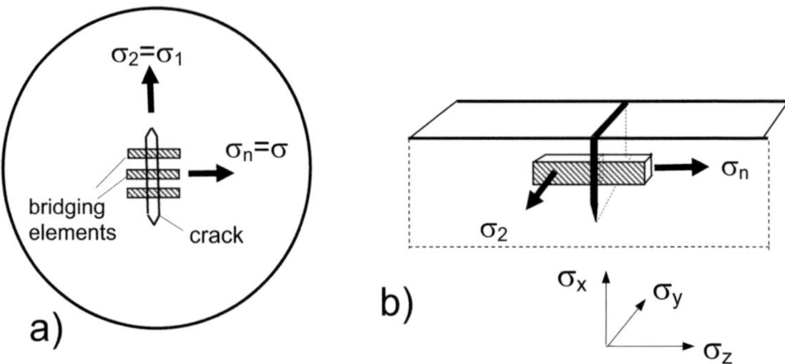

Fig. F2.1 a) Crack in a circular plate under biaxial tension loading (e.g. ring-on-ring test); b) crack with a bridging element under normal tractions $\sigma_n=\sigma_1>0$ and identical tractions parallel to the crack front (coordinates according to Fig. F1.1).

F2.3 A crack under torsion

Figure F2.2 shows a cylindrical specimen loaded by torsion. A crack is assumed to exist with an orientation of 45° with respect to cylinder length axis. This crack is now biaxially loaded by the first principal stress σ_1 acting as the normal traction $\sigma_n=\sigma_1$ on the crack plane and an 'in-plane' stress component $\sigma_3=-\sigma_1$. Also in this case, bridging is affected by the y-stress component parallel to the crack plane. Under a positive normal traction $\sigma_n>0$, the negative σ_3-stress strengthens the bridging interactions

$$\sigma_0 = \sigma_{00}[1 + c(\lambda\sigma_1 - \sigma_3)] = \sigma_{00}[1 + c\sigma_1(\lambda+1)] \tag{F2.3.1}$$

with the result of

$$K_{br} = K_{br,0}[1 + c\sigma_1(\lambda+1)] \tag{F2.3.2}$$

i.e. the bridging effect in torsion is increased.

A similar stress state with different signs for the two principal stresses is present in the Brazilian-disk test ([F2.2], F2.3), F2.4]). In this case it holds $\sigma_2=-3\sigma_1$, i.e. a strong compression loading parallel to the usual crack orientation.

The effect of the multi-axiality of external loading on the R-curves is illustrated schematically in Fig. F2.3. The most important expectations for the frictional contributions to the R-curve are:

- The R-curves are ranked as $K_{R\ (ring\text{-}on\text{-}ring)} < K_{R\ (uniaxial)} < K_{R\ (torsion\ 45°)}$.

- The R-curves from frictional bridging depend on the height of the applied load (K_R depends explicitly on the externally applied load) since the contribution of the clamping mechanism is proportional to the height of the load.

- All multi-axial stress states considered before affect the bridging stress intensity factor although the crack loading is purely mode-I, i.e. a mode-II stress intensity factor does not exist.

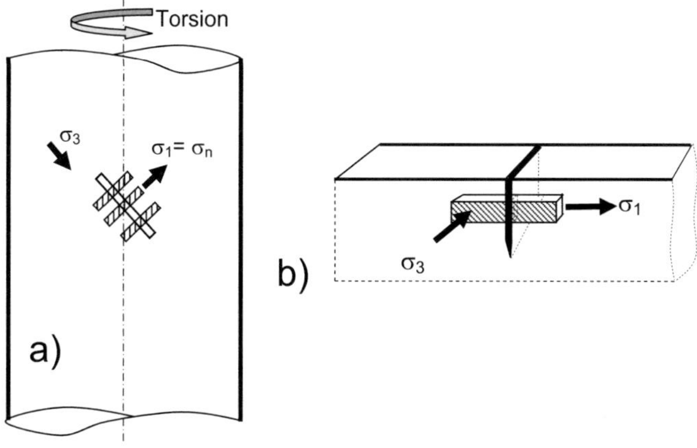

Fig. F2.2 a) Cylindrical specimen under torsion load with a crack at the surface under 45° to the length axis, b) tractions on the bridging element (hatched cuboid).

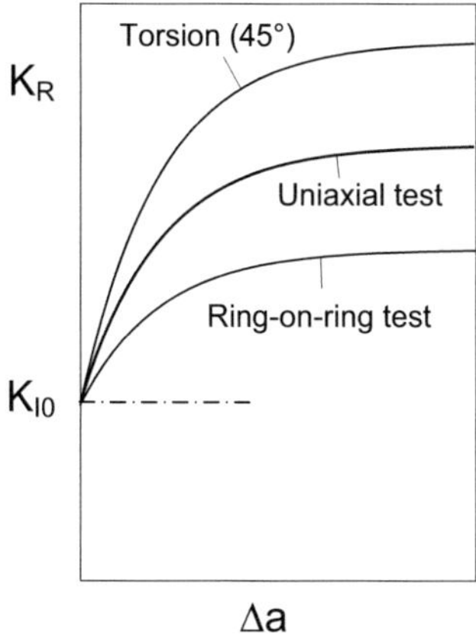

Fig. F2.3 Comparison of the R-curves for the different loading cases at T=0 (schematic).

From the preceding considerations, it may be concluded once more that:

1. The R-curves caused by frictional interface bridging must be affected by multi-axial stresses. Tensile stress components parallel to the crack plane will lower friction effects. Compressive stresses must increase them.

2. R-curves in torsion tests with cracks normal to the first principal stress (i.e. under 45° with respect to the specimen length axis) are expected to be higher than R-curves from uniaxial tests.

3. R-curves from equi-biaxial tests (e.g. ring-on-ring tests or thermally shocked specimens) should be lowered.

References F2

F2.1 Giovan, M.N., Sines, G., Strength of a ceramic at high temperatures under biaxial and uniaxial tension, J. Am. Ceram. Soc. **64**(1981), 68-73.
F2.2 Carneiro, F.L.L.B., Barcellos, A., Résistance à la traction des bétons, Instituto Nacional de Technologia, (1949), Rio de Janeiro.
F2.3 Wright, P.J.F., Comments on an indirect tensile test on concrete cylinders, Magazine of Concrete Research 7(1955), 87–96.
F2.4 Szendi-Horvath, G., Fracture toughness determination of brittle materials using small to extremely small specimens, Engng. Fract. Mech. **13**(1980), 955–961.

F3 Bridging degradation by cyclic loading

F3.1 Frictional bridging degradation

In Section C3, bridging effects during stable crack propagation were considered. The bridging stresses as the true material property could be represented as a function of the actual crack opening displacements. As has been discussed in Sections C3.4, F1 and F2 this statement has to be restricted to the special case of stable crack growth for uni-axial stress and disappearing T-stress. For crack extension under cyclic loading there is much evidence for a reduction of the shielding term K_{sh} with increasing number of cycles. This was found especially for macroscopic cracks. It was very early assumed that the cyclic fatigue behaviour is a direct consequence of the R-curve effect caused by crack surface interactions. The assumption that crack surface interactions become more and more dissolved with increasing number of cycles could be confirmed by in situ microscopic examinations of Frei and Grathwohl [F3.1] and Lathabai et al. [F3.2].

A possible scenario is schematically shown in Fig. F3.1. Figure F3.1a illustrates a beta-crystal orientated normal to the crack surface. As outlined in Section C3.2, this grain will not debond under load application. It elastically bridges the opposite crack-faces, symbolically represented by a spring. Beta-grains under an angle $\Theta<\Theta_{cr}$ are able to debond, F3.1b. At higher load also pull-out will occur resulting in a bending of the beta-grain. Elastic bridging stresses are parallel and in series to the friction tractions.

The frictional crack-face interaction may be described again with the friction model by Mai and Lawn [F3.3]. In Fig. F3.1c, a large grain is shown, acting as a frictional crack-bridging event. The normal component of the thermal mismatch tractions σ_{mis} is indicated.

During crack-face separation resulting in an increasing displacement, δ, a friction stress σ_{fr} occurs which is proportional to the mismatch stress σ_{mis}. The loads transferred by crack face interactions are proportional to σ_{br}, which can be expressed by

$$\sigma_{br} \cong \mu\sigma_{mis} \qquad (F3.1.1)$$

with an *effective* friction coefficient μ.

Due to the repeated sliding of the crack borders under cyclic loading, the contact areas rub on each other. This may lead to a smoothing of surface roughness and a reduction of the friction coefficient with increasing number of cycles [F3.2, F3.4, F3.5]. In addition also friction wear will reduce the transferred forces. As a simple approach, it was proposed in [F3.5]

$$\mu = \mu_0 \exp(-N / N_0) \qquad\qquad\qquad (F3.1.2)$$

where μ_0 stands for the initial friction coefficient. N_0 is a characteristic number of cycles, which of course may depend on the R-ratio and the upper load, since the friction length in each cycle depends on these parameters. Consequently, the crack surface interaction is reduced and the bridging stresses and bridging stress intensity factor are diminished.

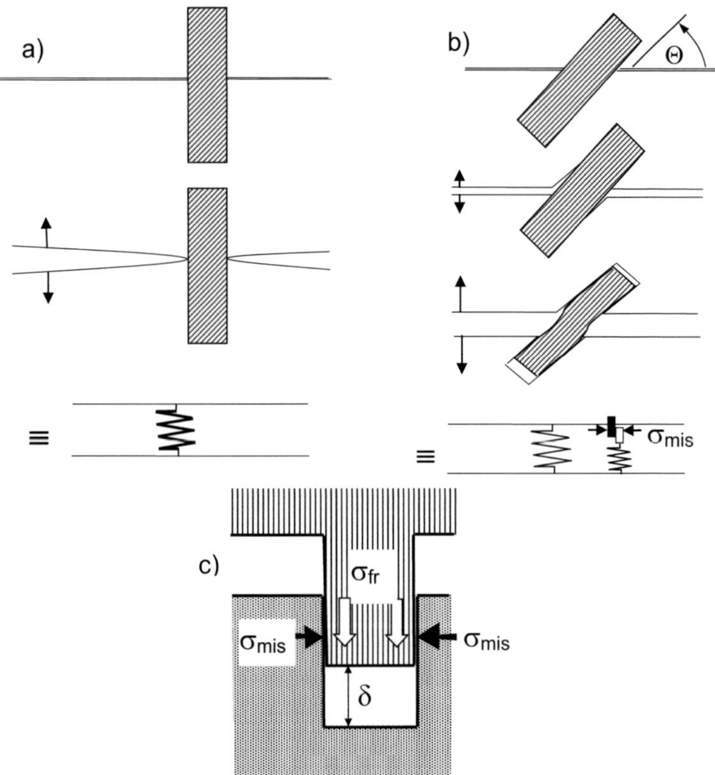

Fig. F3.1 Elastic and frictional crack surface interactions.

F3.2 Crack growth under fatigue loading

The decreasing bridging stress σ_{br} and bridging stress intensity factor K_{br} result in a cycle dependent degradation of the R-curve. This behaviour may be represented as the result of an idealized cyclic test, Fig. F3.2, in which a certain crack length Δa might be generated during the first cycle as indicated in Fig. F3.2a by the circle. This crack is then assumed to be cycled with small loading amplitude at Δa=const with the K_R-degradation according to the solid arrow in Fig. F3.2b. Unavoidable crack propagation predominantly at higher amplitudes results in a curvature of the $K_R(N)$-dependency (dashed arrow).

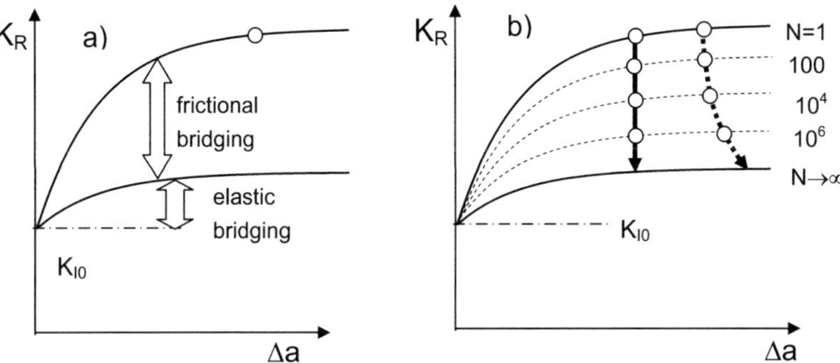

Fig. F3.2 Reduction of the R-curve by degradation of frictional bridges (schematic), a) frictional and elastic bridging stress intensity factor contributions, b) influence of number of cycles.

Whereas the frictional bridging effects decrease via a reduction of μ, the elastic bridges can fail by quasi-static subcritical crack growth. This means, the time scale for failure by quasi-static subcritical crack growth is different from that for cyclic fatigue, i.e. it has to be expected that the fatigue effect on frictional and elastic bridges under cyclic loading may occur on different time (or cycle) scales. For reasons of clearness, a reduction of elastic bridges is neglected in the following considerations.

The R-curve is schematically plotted in Fig. F3.2b for an increasing number of cycles N. The quasi-static R-curve is represented by N=1. In our model, only the frictional bridges are reduced. The curve for N→∞ then reflects the R-curve effect of the remaining elastic bridges. In practice, also the value of K_{I0} may be reduced during cycling especially in presence of quasi-static subcritical crack growth enhanced in water or humid air. This additional effect, obvious from R-curve measurements by Kruzic et al. [F3.6, F3.7], is ignored in Fig. F3.2b for reasons of simplicity.

F3.2.1 Fatigue crack growth in K-controlled tests

The fatigue effect by degradation of the R-curve is illustrated in Fig. F3.3. Here, the special case of a K-controlled test is assumed. Such a test is of course only practicable for macroscopic cracks. Figure F3.3a shows the stress intensity factor vs. time history with the maximum and minimum stress intensity factors, K_{max} and K_{min}, kept constant. The actual crack length a is defined by the intersection of K_{max} with the time-dependent R-curve (Fig. F3.3b). The crack lengths after $N=1$, 10^4, and 10^6 cycles are indicated by the circles. Failure of the specimen in this case would occur after about $N\approx10^7$-10^8 cycles because no intersection with the curve for $N\rightarrow\infty$ exists.

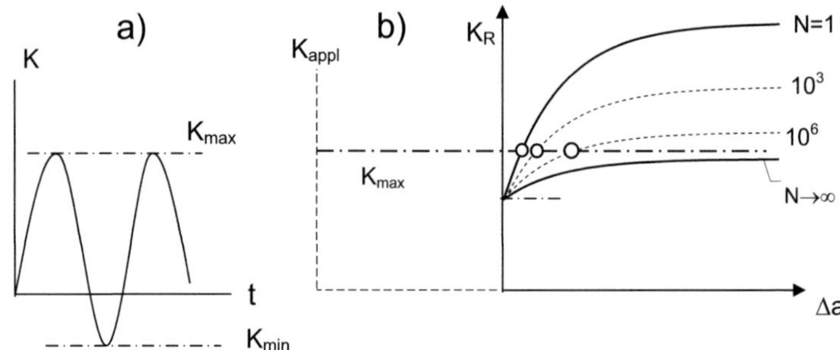

Fig. F3.3 Fatigue crack growth due to degradation of the R-curve, a) load history, b) crack extension with number of cycles.

The actual crack growth rates are given by

$$\frac{da}{dN} = \underbrace{\left.\frac{da}{dK_R}\right|_{K_{appl}}}_{\substack{\text{reziprocal steepness}\\\text{of the R-curve}}} \underbrace{\frac{dK_R}{dN}}_{\text{degradation}} \tag{F3.2.1}$$

where the left-hand side is a measurable quantity.

If the upper load K_{max} is smaller or if the R-curve for $N\rightarrow\infty$ will monotonously increase with crack length, the effect of crack arrest should be observed. Figure F3.4 illustrates this situation. For an experimental procedure to determine the fatigue-R-curve as the R-curve for $N\rightarrow\infty$ and experimental results, see Kruzic et al. [F3.7].

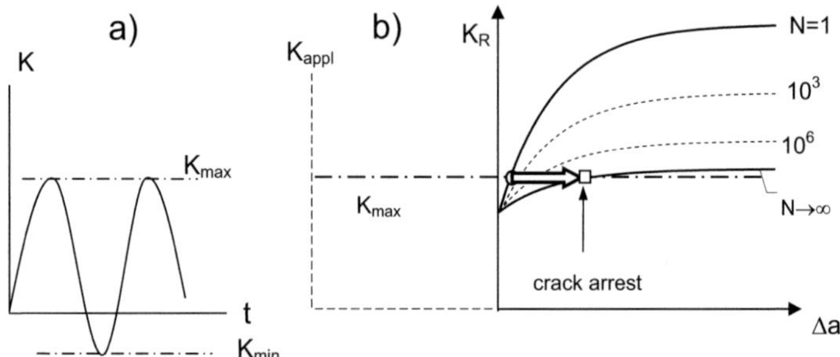

Fig. F3.4 Crack arrest for a test with a small K_{max}.

F3.2.2 Fatigue crack growth in stress-controlled tests

In fatigue tests, mostly specimens with natural crack population are used. Such experiments are performed stress-controlled with the externally applied stresses kept constant during the whole test. The applied stress intensity factor, therefore, increases monotonously with time since the crack length a increases with number of cycles (Fig. F3.5a).

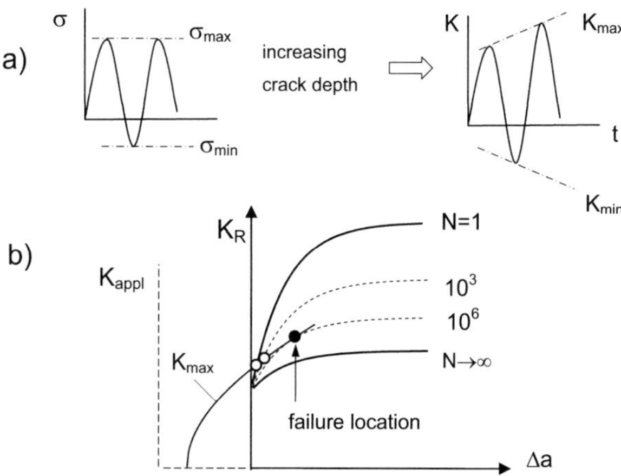

Fig. F3.5 Fatigue crack growth in stress-controlled tests as usual for specimens with natural crack population, a) load history, b) crack extension for an increased number of cycles.

Also in this test, the actual crack length a is defined by the intersection of the roughly square-root shaped curve for $K_{max}(a)$ with the cycle-dependent R-curve (Fig. F3.5b). The crack lengths after $N = 1$ and 10^3 cycles are indicated by the open circles. At $N = 10^6$ the tangent condition is fulfilled (indicated by the solid circle) and the specimen fails.

F3.3 A fatigue-R-curve for natural cracks

The failure condition (illustrated in Fig. F3.5b) allows a fatigue R-curve for natural cracks to be derived. For this purpose, it is necessary to determine the static R-curve, the inert strength σ_c, and the SN-diagram with a sufficiently large number of tests on each stress level. From these data, a special type of fatigue R-curve for a constant number of cycles can be constructed as follows. Figure E3.6 illustrates schematically the results for the inert strength and the number of cycles to failure, both in Weibull representation. In Fig. E3.6a the inert strength σ_c is plotted versus the cumulative failure probability F. Figure E3.6b shows the Weibull distributions for N_f obtained at 3 different upper stresses σ_{max}.

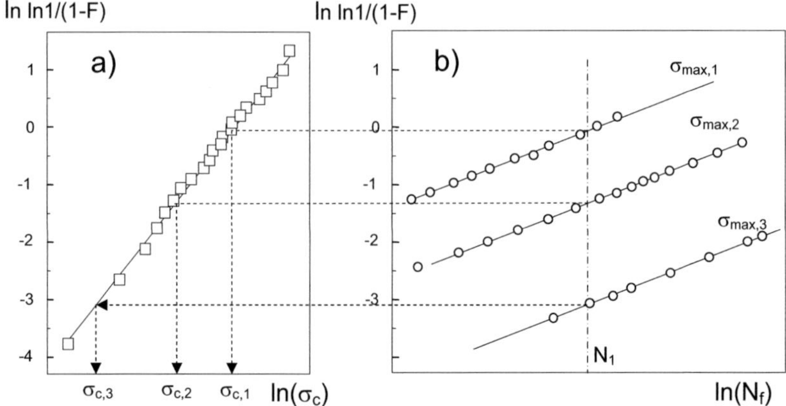

Fig. E3.6 Schematic of strength and cycles-to-failure data, a) scatter of inert strength represented in a Weibull diagram, b) Weibull plot of number of cycles to failure for 3 load levels (immediate failure and survivals not potted).

The determination of the cyclic fatigue R-curve may be explained for the number of cycles N_1. From the Weibull plot Fig. E3.6b we get the failure probabilities for N_1 as the intersections of the Weibull lines with the dash-dotted line. From the Weibull curve of the inert strength, Fig. E3.6a, we find the inert strengths for the same failure probabilities (dashed lines). These values are denoted as $\sigma_{c,1}$ ... $\sigma_{c,3}$. Now pairs of corresponding strengths and numbers of cycles to failure ($\sigma_{c,i}$, $\sigma_{max,i}$) are available.

We now assume that the R-curve for the fictive material of Fig. E3.6 might be also available. Figure E3.7 shows the R-curve as the solid line. Since the strengths $\sigma_{c,1}$ ... $\sigma_{c,3}$ are the applied stresses for which the tangent conditions between the applied stress intensity factors $K_{appl,i}$ and the K_R-curve is fulfilled, the initial crack lengths $a_{0,i}$ result from the solution of

$$K_{appl} = K_{IR} \tag{F3.3.1}$$

$$\frac{dK_{appl}}{da} = \frac{dK_{IR}}{da} \tag{F3.3.2}$$

with
$$K_{appl} = \sigma Y \sqrt{a}, \quad Y \cong 1.3, \tag{F3.3.3}$$

This can be done graphically as schematically shown in Fig. F3.7 for two of the strength values. In practice, it is of advantage to fit the initial part of the R-curve and to evaluate the tangent condition analytically.

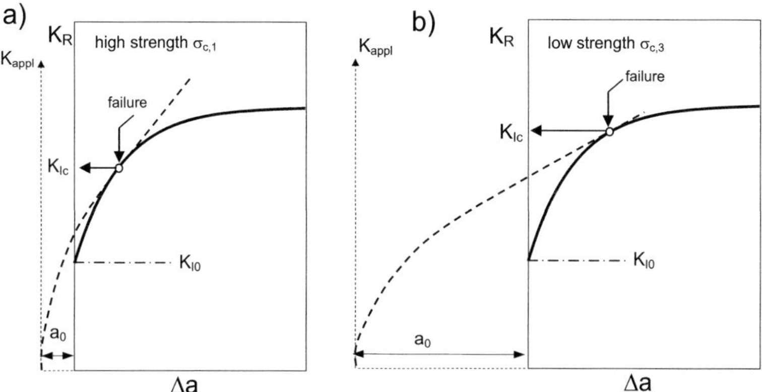

Fig. F3.7 Determination of the initial crack size a_0 from the tangent condition.

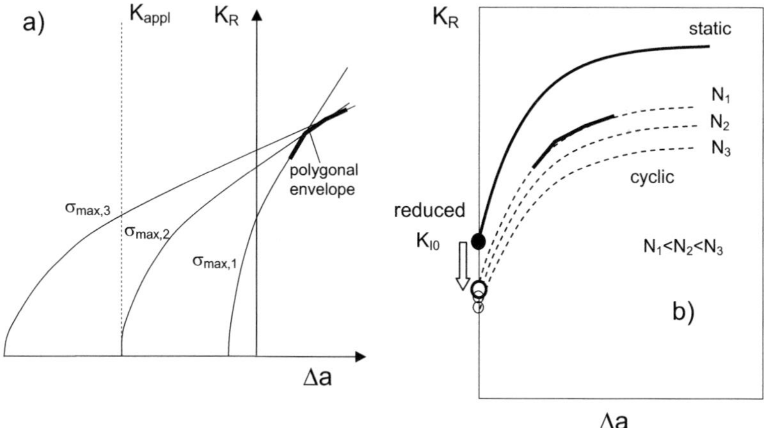

Fig. F3.8 Determination of the fatigue R-curves for natural cracks.

Next, we have to take into account that failure in the fatigue tests takes place at a lower applied load $\sigma_{max,i} < \sigma_{c,i}$. By use of the three $\sigma_{max,i}$-values, eq.(F3.3.3) yields a series of 3 K_{appl} curves. The fatigue R-curve must touch each of the single K_{appl} curves in one point (where the failure condition in the fatigue test is fulfilled) and otherwise must lie below the envelope established by the K_{appl}-curves (Fig. F3.8a). The so-constructed bold polygonal curve represents an upper limit of the fatigue R-curve in a region relevant for fatigue failure. If we smooth this upper limit curve at the crossing points of the individual K_appl-curves, an improved approximation is obtained. Since $\sigma_{\text{max},i} < \sigma_{c,i}$, the fatigue R-curve is less high than the static R-curve. Together with a possibly reduced K_{I0} an extended curve can be derived (Fig. F3.8b). It has to be emphasized that this curve is correct only in the region of the crossing K_appl-curves. The extended curve between this region and K_{I0} represents an approximation only.

The procedure described before has to be performed for different numbers of cycles to failure N_i as indicated in Fig. F3.8b.

By using the extended (dashed) curves, the lifetime prediction of cracks with different initial sizes is possible.

F3.4 Cyclic fatigue at high temperatures

Let us now address the temperature influence on fatigue behaviour by consideration of two limit temperatures. These limits are room temperature (RT) at which the maximum thermal mismatch appears and sintering temperature (ST) where any thermal mismatch must disappear. Figure F3.9a again shows the frictional part of the bridging stress intensity factor, $K_\text{br,frict}$, as a function of numbers of cycles and temperature as the open arrow and the nearly cycle-independent elastic bridging part, $K_\text{br,elast}$, as the full arrow. Due to the lack in thermal mismatch stresses and, consequently, in frictional bridges at sintering temperature, a cyclic fatigue effect by degradation of frictional bridges might be not observable. In this approximation specimens loaded at ST will either fail spontaneously if $K_\text{appl} \geq K_{I0} + |K_\text{br,elast}|$ or will survive a fatigue test if $K_\text{appl} < K_{I0} + |K_\text{br,elast}|$.

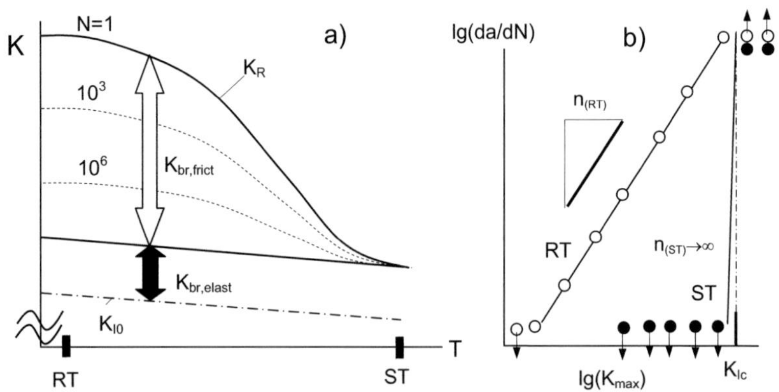

Fig. F3.9 Effect of temperature on cyclic fatigue exponent n; a) decrease of the frictional bridging stress intensity factor for reduced thermal mismatch at increased temperature, b) effect on the da/dN-K_max curves (RT: room temperature, ST: sintering temperature).

An evaluation of the cyclic lifetime tests in order to determine $da/dN=f(K_{max})$-curves must result in increasing steepness of curves with a decreasing number of tests with failure in a certain finite test duration. This is identical with an increasing exponent n with increasing temperature in the power-law representation

$$\frac{da}{dN} = AK_{max}^{n} \qquad\qquad (F3.4.1)$$

At sintering temperature, it must hold $n_{(ST)}\rightarrow\infty$. This behaviour is illustrated in Fig. F3.9b. The circles may indicate the individual measurements. The upward arrows symbolize tests resulting in spontaneous failure, the downward arrows tests which did not fail within a prescribed time span. The slopes of the straight lines through the results with finite lifetimes are identical with the exponent n in eq.(F3.4.1).

For reasons of clearness, the simplified description in this section did not take into account possible interactions of frictional and elastic bridging effects. Such an interaction is for instance the increase of load that has been transferred by an elastic bridge when the frictional bridging events continuously decrease their load capability. This can lead to a delayed failure of an elastic bridge by reaching its strength without showing a true fatigue effect.

F3.5 Interaction of cyclic and quasi-static subcritical crack growth

As already mentioned in context with a reduced K_{I0}-value in presence of humidity, the effect of subcritical crack growth will be superimposed to the "true" fatigue crack growth by cyclic bridging degradation. This can easily be seen from fatigue experiments on coarse-grained alumina with constant R-ratio but different frequency f [F3.8]. In the case of a real fatigue effect, the lifetime must decrease with increasing frequency. Three series of specimens were cyclically loaded with loudspeakers by use of a cantilever arrangement. Alternating bending ($R=-1$) was chosen as the loading.

Figure F3.10a shows the lifetimes obtained at a maximum stress of $\sigma_{max} = 175$ MPa and frequencies of 0.2, 2 and 20 Hz. As can be seen, the lifetimes decrease with increasing frequency. This is an indication that not only subcritical crack growth may be the reason for failure in cyclic tests.

There is a clear trend of increasing lifetime with decreasing frequency. Figure F3.10b represents the results of Fig. F3.10a with the lifetime replaced by the number of cycles to failure. From Fig. F3.10b we can conclude that also cycle effects are not the only reason for crack extension, because the numbers of cycles to failure do not coincide for the different frequencies. Consequently, we have to conclude that for alumina the degradation effect <u>and</u> subcritical crack growth are both present in fatigue tests carried out in air.

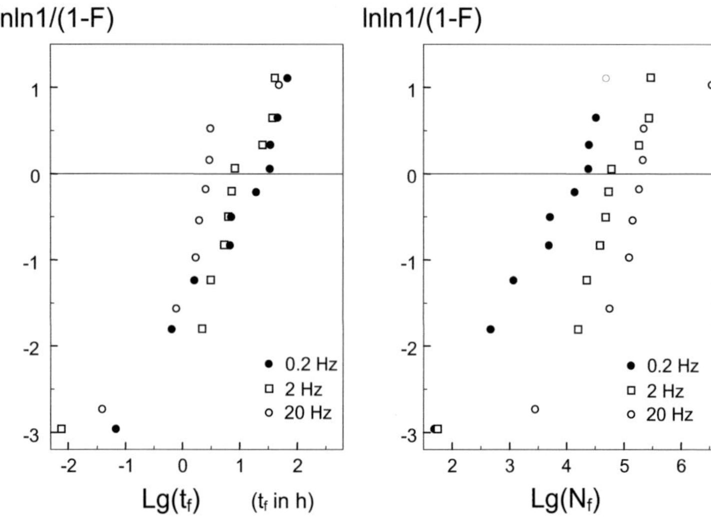

Fig. F3.10 a) Lifetimes under alternating bending load ($R = -1$) with maximum stress $\sigma_{max} = 175$ MPa, measured for different frequencies of 0.2 Hz, 2 Hz and 20 Hz; b) the same results represented as a function of the number of cycles to failure.

References F3

F3.1 Frei, H., Grathwohl, G., The fracture resistance of high performance ceramics by in situ experiments in the SEM, Beitr. Elektronenmikroskop. Direktabb. Oberfl. **22**(1989), 71–78.

F3.2 Lathabai, S., Rödel, J., Lawn, B.R., Cyclic fatigue from frictional degradation at bridging grains in alumina, J. Am. Ceram. Soc. **74**(1991), 1340–48.

F3.3 Mai, Y., Lawn, B.R., Crack-interface grain bridging as a fracture resistance mechanism in ceramics: II. Theoretical fracture mechanics model, J. Am. Ceram. Soc. **70**(1987), 289.

F3.4 Dauskardt, R.H., Acta metal. Mater. **41**(1993), 2765.

F3.5 T. Fett, O. Kraft, D. Munz, Fatigue failure of coarse-grained alumina under contact loading, Mat.-wiss. u. Werkstofftech. **36**(2005), 163-170.

F3.6 S. Gallops, T. Fett, J.J. Kruzic, Fatigue threshold R-curve behaviour of grain bridging ceramics: Role of grain size and boundary adhesion, J. Am. Ceram. Soc. (2011), in press.

F3.7 J.J. Kruzic, R.M. Cannon, J.W. Ager III, R.O. Ritchie, Fatigue threshold R-curves for predicting reliability of ceramics under cyclic loading, Acta Materialia **53**(2005), 2595-2605.

F3.8 T.Fett, D. Munz, G. Thun, Influence of frequency on cyclic fatigue of coarse-grained Al_2O_3, J. Mater. Sci. Letters **12**(1993), 220-22.

G1 Shielding effects under mode-II loading

G1.1 Mode-II shielding due to crack-face roughness

G1.1.1 Shear stresses

Real crack surfaces are not ideally straight in a mathematic sense. This is schematically illustrated in Fig. G1.1a. Under pure shear loading, τ, the upper and lower crack faces are shifted by the shear displacements δ_{II}. Due to the interlocking of the two crack faces, this shift causes friction-like shear tractions τ_{fr} acting against the externally applied stresses τ_{appl}. The total shear results as

$$\tau_{total} = \tau_{appl} + \tau_{fr} \qquad \text{(G1.1.1)}$$

In addition also a normal displacement δ_I results even under pure shear as indicated in Fig. G1.1b. Consequently, the frictional contact undergoes a normal pressure, which would be necessary in a mode-I loading for evoking the same displacement δ_I. This crack opening gives rise for a small mode-I stress intensity factor K_I also in the absence of externally applied normal tractions on the crack surface as outlined by Mendelsohn et al. [G1.1, G1.2]. This additional mode-I stress intensity factor is in most cases small compared to the usually applied one.

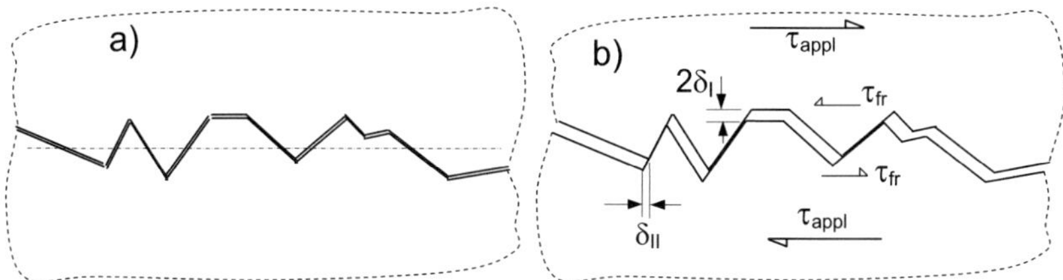

Fig. G1.1 Crack opening displacements under externally applied shear stresses (heights exaggerated), a) without externally applied tractions, b) pure shear applied.

G1.1.2 Frictional stress intensity factor

If the distributions of the shear stress contributions in eq.(G1.1.1) are know, the weight function method allows the related stress intensity factor parts to be computed. The applied mode-II stress intensity factor is given by

$$K_{II,appl} = \int_0^a \tau_{appl} h_{II}(a,x)\,dx \tag{G1.1.2}$$

where h_{II} is the fracture mechanics mode-II weight function and x the crack coordinate with origin at the free surface. For a constant shear $\tau_{appl} \neq f(x)$, it simply holds with the geometric function F_{II}

$$\int_0^a \tau_{appl} h_{II}(a,x)\,dx = F_{II}\sqrt{\pi a} \tag{G1.1.3}$$

Similar to eq.(G1.1.2), the frictional and the total mode-II stress intensity factors can be written as

$$K_{II,fr} = \int_0^a \tau_{fr} h_{II}(a,x)\,dx \tag{G1.1.4}$$

$$K_{II,total} = \int_0^a \tau_{total} h_{II}(a,x)\,dx \tag{G1.1.5}$$

resulting in

$$K_{II,total} = K_{II,appl} + K_{II,fr} \tag{G1.1.6}$$

Equation (G1.1.4) gives rise for the definition of an effective constant shear stress τ_{eff}. Applying the mean value theorem for integrals on eq.(G1.1.4) yields

$$K_{II,fr} = \int_0^a \tau_{fr} h_{II}(a,x)\,dx = \tau_{fr,eff} \int_0^a h_{II}(a,x)\,dx = \tau_{fr,eff} F_{II}\sqrt{\pi a} \tag{G1.1.7}$$

G1.1.3 Fatigue effect under cyclic mode-II loading

Similar to fatigue of frictional crack-face interactions addressed in Section F3, also the effective frictional shear must decrease with number of cycles. This is schematically illustrated in Fig. G1.2a. The fatigue behaviour is mostly interpreted as a wear and polishing effect during periodical crack face sliding [G1.3, G1.4]. Under cyclic mode-II loading, the crack faces rub on each other. This may lead to an increasing smoothing of surface roughness and a reduction of the friction coefficient with increasing number of cycles as was successfully assumed for the interpretation of contact fatigue experiments [G1.5].

Figure G1.2b illustrates the frictional tractions near a crack tip and its degradation with increasing number of cycles N.

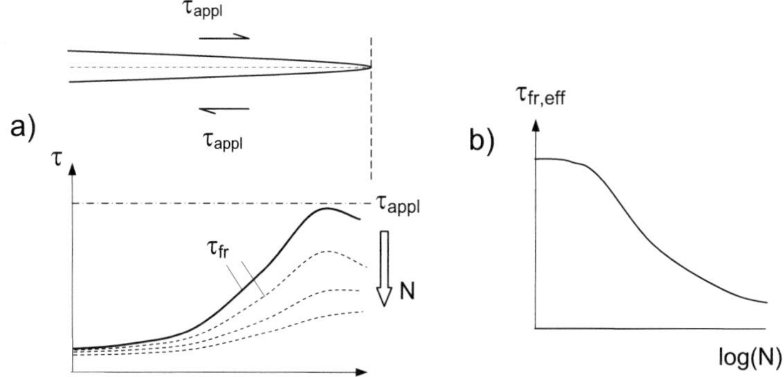

Fig. G1.2 a) Distribution of the frictional shear tractions caused by crack-face roughness; arrow indicates friction degradation by cyclic shear loading, b) degradation of the effective frictional shear tractions.

G1.2 Crack-surface bridging under mixed-mode loading

G1.2.1 Mode-II shielding by bridging under pure shear

In the same way as roughness can interlock the two opposite faces of a crack [G1.1-G1.3], it has to be expected that crack-face interactions via grains will also affect crack extension under pure or superimposed mode-II loading as schematically illustrated in Fig. G1.3. In [G1.6] it was outlined that shear tractions and a mode-II stress intensity factor K_{II} are generated under small mode-II load contributions as occur for instance by small misalignments Δ in the load application.

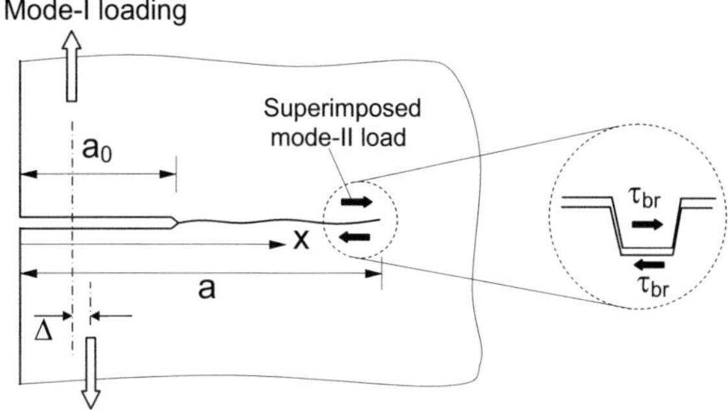

Fig. G1.3 Mode-II load generated by a misalignment Δ in the mode-I load application.

The mode-II stress intensity factor gives rise for displacements parallel to the crack plane. Due to a crack-face interlocking by single grains, the crack tip is shielded from the mode-II load by tangentially transferred tractions τ_{br} as illustrated in Fig. G1.3 by the dashed circle.

The mode-II shielding stress intensity factor, $K_{II,sh}$, can be computed from the distribution of the shear tractions τ_{br} along the crack. It holds by use of the mode-II weight function h_{II}

$$K_{II,sh} = \int_{a_0}^{a} \tau(x)\, h_{II}(x)\, dx \tag{G1.2.1}$$

The actual mode-II crack tip stress intensity factor $K_{II,tip}$ results from

$$K_{II,tip} = \begin{cases} 0 & for \quad |K_{II,appl} + K_{II,sh}| \le 0 \\ K_{II,appl} + K_{II,sh} & else \end{cases} \tag{G1.2.2}$$

This stress intensity factor governs the local stability of crack paths. If the value $K_{II,tip}$ does not disappear, the crack must kink out of the initial crack plane (see Section G.2). This effect is highly undesired in mechanical material testing.

G1.2.2 Mixed-mode shielding by crack-face bridging

First measurements on K_{IIR} were reported in [G1.7] but not evaluated with respect to the parameters of a special model. A commercial alumina with a median grain size of $d_m \approx 9$ µm was tested. Specimens of $3\times4\times45$ mm^3 were loaded by the force P via two opposite spheres of radius $R=5$mm (Fig. G1.4a). Tests under monotonously increasing load were performed up to varying maximum loads. The specimens were cut through the centre of the contact area. Under the SEM, the main crack dimensions were measured. For the cone angles an average value of $\alpha=31°$ was found.

In all cases straight cracks were found indicating that always $K_{II,tip}=0$ was fulfilled in the tests. Together with eq.(G1.2.2) this condition yields a simple relation for the computation of the mode-II shielding term $K_{II,sh}$, namely

$$K_{II,sh} = -K_{II,appl} \tag{G1.2.3}$$

Stress intensity factor computations by a 3D finite element analysis were performed in [G1.8], providing $K_{I,appl}$ and $K_{II,appl}$.

For the angle of $\alpha \cong 31°$ a significant mode-II stress intensity factor contribution $K_{II,appl}$ was obtained.

Figure G1.4b shows the mode-I shielding stress intensity factor data determined from the microscopically measured crack lengths a, the applied load P and the result of the finite element computations. It has to be considered that the scatter in data is caused by the application of a multi-specimen test and not the result of one single propagating crack at different crack lengths.

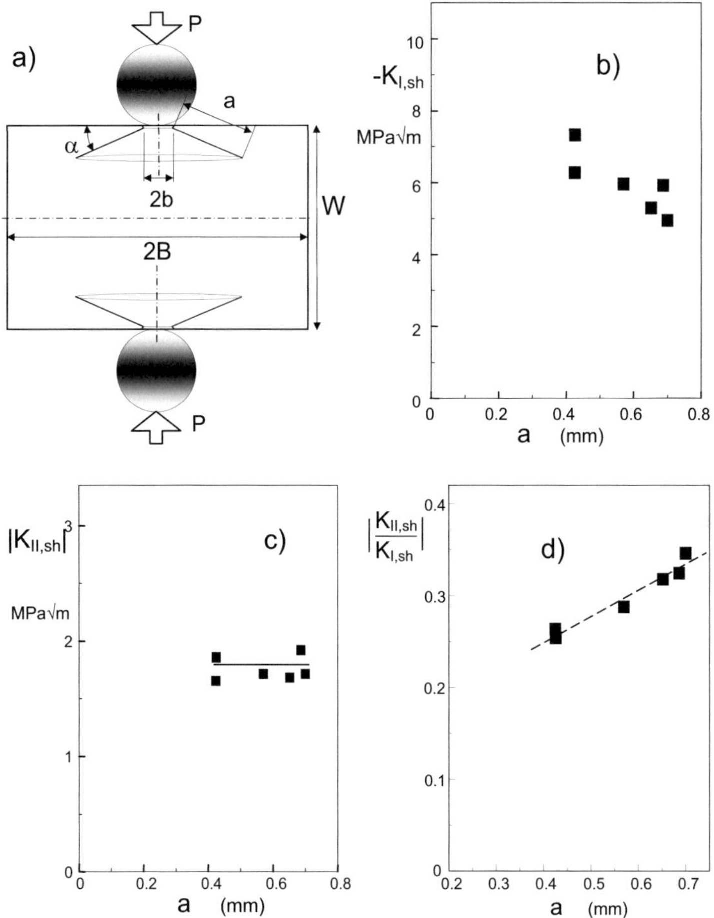

Fig. G1.4 a) Opposite cone cracks in a rectangular bar (geometric data), b) mode-I shielding stress intensity factor $K_{I,sh}$, c) mode-II shielding term $K_{II,sh}$, d) ratio of the two shielding contributions.

In the same way also the shielding stress intensity factor $K_{II,sh}$ at the different crack lengths could be determined. Figure G1.4c shows the results.

Finally, the individual values of the ratio $K_{II,sh}/K_{I,sh}$ are given in Fig. G1.4d. The ratio of the two shielding terms is roughly 0.3 but shows an increase with increasing crack length.

References G1

G1.1 Mendelsohn, D.A., Gross, T.S., Zhang, Y., Acta metall mater. **43**(1995), 893-900.

G1.2 Mendelsohn, D.A., Gross, T.S., Goulet, R.U. Zhouc, M., Materials Science and Engineering **A249**(1998), 1- 6.

G1.3 Lathabai, S., Rödel, J., Lawn, B.R., J. Am. Ceram. Soc. **74**(1991), 1340.

G1.4 Dauskardt, R.H., Acta. Metal. Mater **41**(1993), 2765.

G1.5 Fett, T., Kraft, O., Munz, D., Fatigue failure of coarse-grained alumina under contact loading, Mat.-wiss. u. Werkstofftech. **36**(2005), 163-170.

G1.6 Fett, T., Rizzi, G., Munz, D., Hoffmann, M., Oberacker, R., Wagner, S., Bridging interactions in ceramics and consequences on crack path stability, J. Ceram. Soc. Japan, **114**(2006), 1038-1043

G1.7 Fett, T., Rizzi, G., Munz, D., Badenheim, D., Oberacker, R., Sphere contact fatigue of a coarse-grained Al_2O_3 ceramic, Fatigue&Fracture of Engineering Materials and Structures, **29**(2006), 867.

G1.8 T. Fett, G. Rizzi, M. Hoffmann, R. Oberacker, S. Wagner, Mode-II shielding-curve of Al_2O_3 from measurement of cone crack angles, J. Mater. Sci. **43**(2008), 2077-2081.

G2 R-curve and path stability

G2.1 Conditions for path stability

A crack of initial length a_0 is considered, which grows out of the initial straight plane by an angle of Θ_0 (Fig. G2.1a). The kink angle Θ_0 may represent the influence of a disturbing mode-II loading caused by a small unavoidable misalignment of the loading arrangement. For small kink angles it holds

$$\Theta_0 = -\frac{2K_{\mathrm{II}}}{K_{\mathrm{I}}} \tag{G2.1.1}$$

where K_{I} and K_{II} are the stress intensity factors for the initial crack situation.

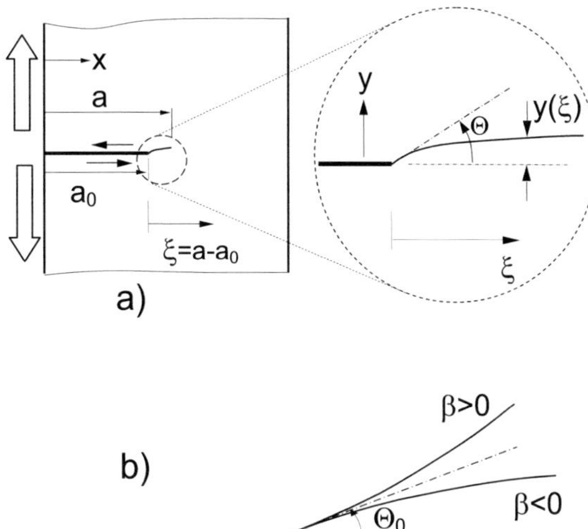

Fig. G2.1 a) Geometrical data of a crack growing under mode-I loading (vertical arrows) with a superimposed small mode-II disturbance (horizontal arrows), b) general influence of the T-stress after crack kinking under mixed-mode loading.

In [G2.1], Cotterell and Rice analyzed the local crack path stability where the first deviation from the initial straight crack plane was of interest. At small extensions ξ, the deviation from the initial crack plane, y, is given by

$$y(\xi) = \frac{\Theta_0}{\beta^2} \frac{a_0 \pi}{8} \left[\exp\left(\frac{8\chi}{\pi}\beta^2\right) \mathrm{erfc}\left(-\beta\sqrt{\frac{8\chi}{\pi}}\right) - 1 - \frac{4}{\pi}\beta\sqrt{2\chi} \right]$$

$$\chi = (a - a_0)/a_0$$

(G2.1.2)

with the biaxiality ratio β (see Section B1.3).

Equation (G2.1.2) is a solution for short crack extensions. This relation allows to discuss the basic influences on path stability.

The most important conclusion of [G2.1] is illustrated in Fig. G1.1b, namely, increasing deviation from the prescribed kink angle for $\beta>0$ and decreasing deviations for $\beta<0$.

From eq.(G2.1.2), it has to be expected that crack path stability is only guaranteed for $\beta<0$. In nearly all fracture mechanics test specimens, however, the T-stress terms and, consequently, the biaxiality ratios β are positive, at least in the commonly used range of crack lengths, Fig. G2.2a.

There are two exceptions for standard test specimens used for ceramics, namely, small cracks in bending bars with a relative crack length $a/W<0.35$ (specimen width W) and the DCDC specimen that shows strongly negative β in the whole range of possible crack lengths [G2.2] (see Fig. G2.2b).

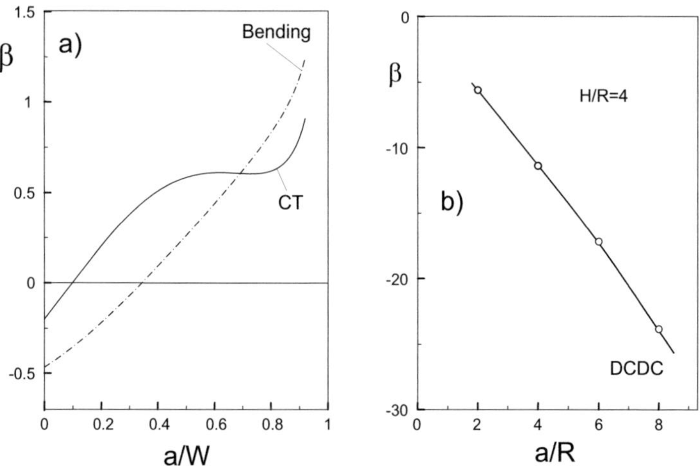

Fig. G2.2 a) Biaxiality ratio for 4-point bending and compact tension (CT) specimens, b) for the DCDC specimen.

From the mostly positive sign of the biaxiality ratio for test specimens and the unavoidable small misalignments, it has to be expected that path stability during crack growth would be impossible.

Nevertheless, innumerable experimental results on R-curves are reported in literature. In contrast to the expectation from eq.(G2.1.2), however, no crack-path instability worth mentioning was detected. This is not astonishing for DCDC tests and bending bars with short cracks because of their negative or moderately positive parameter β. Steinbrech et al. [G2.3], for example, measured R-curves on coarse-grained alumina in bending up to relative crack lengths of about $a/W=0.95$ (see Figs. A2.2 and A2.5). For such deep cracks strong path instability has to be expected.

For tests on materials without an R-curve (e.g. DCB tests on glass), it is well known from literature that guiding grooves are indispensable.

For coarse-grained materials, a shielding stress intensity factor exists that shields the crack tip partially from the applied loads as was outlined in Section C3 for mode-I shielding.

G2.2 Influence of bridging interactions on path stability

Let us assume a sharp crack of length a in a specimen, introduced by stable crack propagation starting from a notch of length a_0. During further extension of this crack, a slight misalignment of the load application is assumed, resulting in a mode-II stress intensity factor $K_{II,appl}$. During crack growth under an applied mode-I stress intensity factor $K_{I,appl}$, it holds in the absence of subcritical crack growth that

$$K_{I,tip} = K_{I,appl} + K_{I,br} = K_{I0} \tag{G2.2.1}$$

In the absence of any crack-face interlocking, the applied mode-II loading causes shear displacements $\delta_{x,appl}$ between the two crack surfaces. In the presence of bridging effects with crack-face interlocking, these displacements are (at least partially) suppressed and generate locally concentrated shear forces at the location of the bridging interactions acting in the opposite direction. Similar to the treatment of mode-I loading, a continuously distributed shear stress averaging the local crack face interactions is assumed for the bridging effect under mode-II loading.

The material will exhibit a mode-II bridging stress intensity factor under mode-II loading here denoted by $K_{II,br}$. The total mode-II crack tip stress intensity factor $K_{II,tip}$ results in the same way as discussed for mode-I by superposition of the applied and the bridging (or shielding) stress intensity factors

$$K_{II,tip} = K_{II,appl} + K_{II,br} \tag{G2.2.2}$$

The crack kink angle Θ (in contrast to eq.(G2.1.1)) is given by the ratio of the crack-tip stress intensity factors

$$\Theta = -2\frac{K_{II,tip}}{K_{I0}}$$
(G2.2.3)

As long as the crack-face interactions can carry the mode-II load (upper part of eq.(G1.2.2)), i.e. as long as $|K_{II,br}+K_{II,appl}|\leq 0$, it results for the total stress intensity factor at the crack tip

$$K_{II,tip} = 0$$
(G2.2.4)

and, consequently, $\Theta=\Theta_0=0$, $\Rightarrow y(\xi)=0$. With other words: Due to the crack-face bridging events, crack-path stability is ensured even under rather strong misalignments (for details see also [G2.4]).

References G2

G2.1 Cotterell, B., Rice, J.R., Slightly curved or kinked cracks, Int. J. Fract., **16**(1980), 155-169.
G2.2 Fett, T., Rizzi, G., Munz, D., T-stress solution for DCDC specimens, Engng. Fract. Mech., **72**(2005), 145-149.
G2.3 Steinbrech, R., Reichl, A., Schaarwächter, W., R-curve behaviour of long cracks in alumina, J. Am. Ceram. Soc. **73**(1990), 2009–2015.
G2.4 Fett, T., Rizzi, G., Munz, D., Hoffmann, M., Oberacker, R., Wagner, S., Bridging interactions in ceramics and consequences on crack path stability, J. Ceram. Soc. Japan, **114**(2006), 1038-1043

H1 Phase transformation and micro-cracking

H1.1 Transformation zone and R-curve in zirconia ceramics

Due to the singular stress field near a crack tip in transformation-toughened zirconia, the material undergoes a stress-induced martensitic transformation and the tetragonal phase changes to the monoclinic phase. This transformation occurs when the characteristic local stress σ_{char} reaches a critical value $\sigma_{\text{char,c}}$. The result is a crack-tip transformation zone.

Several stress criteria for the onset of phase transformation were applied in literature. In one of the earliest attempts [H1.1], it was assumed that volume strains of the phase transformation only are playing a part in the transformation criterion, because the transformation shear strains are nearly annihilated by twinning. Since volume strains are assumed to be triggered by the hydrostatic stress σ_{hyd}, a hydrostatic transformation criterion was proposed in [H1.1]

$$\sigma_{hyd} = \tfrac{1}{3}(\sigma_x + \sigma_y + \sigma_z) = \sigma_{hyd,c} \tag{H1.1.1}$$

For the special case of small-scale transformation conditions (transformation zone size negligible compared to crack size and component dimensions), McMeeking and Evans [H1.1] computed the transformation zone, neglecting the perturbation of the stress field due to transformations (modified by Budiansky et al. [H1.2]).

To the knowledge of the author, Giannakopoulos and Olsson [H1.3] published the first theoretical study of the effect of T-stress on phase transformation zones. This investigation is the basis of the following considerations.

In the presence of a T-stress contribution, the *hydrostatic stress* near the tip of the crack under plane strain conditions reads

$$\sigma_{hyd} = \frac{1+v}{3}\left(\frac{2K_{\mathrm{I}}}{\sqrt{2r\pi}}\cos(\varphi/2) + T\right) \tag{H1.1.2}$$

From (H1.1.1) and (H1.1.2), the shape $r(\varphi)$ of the phase transformation zone for plane strain results as

$$r = \frac{8}{3\sqrt{3}}\omega\cos^2(\theta/2) \tag{H1.1.3}$$

with the height ω of the zone

$$\omega = \frac{(1+v)^2}{4\sqrt{3}\pi}\left(\frac{K_{\mathrm{I}}}{\sigma_{\mathrm{hyd,c}} - (1+v)T/3}\right)^2 \qquad\qquad \text{(H1.1.4)}$$

(for r, φ, and ω see Fig. H1.1). Figure H1.1a illustrates the transformation zone for a non-extending crack, Fig. H1.1b the shape after a crack extension of Δa. Due to the martensitic transformation, a volumetric expansion strain of about 4.5% occurs. These strains cause tensile stresses at a certain distance ahead of the crack tip and compressive stresses along the length Δa at the crack line. The compressive stresses lead to a shielding stress intensity factor, which has to be overcome during crack propagation, i.e. the applied stress intensity factor must be increased to maintain stable crack growth.

In later, more complicated numerical studies (e.g. [H1.4]) the influence of shear stresses and shear strains on the transformation criterion and on the zone size and shape was taken into consideration.

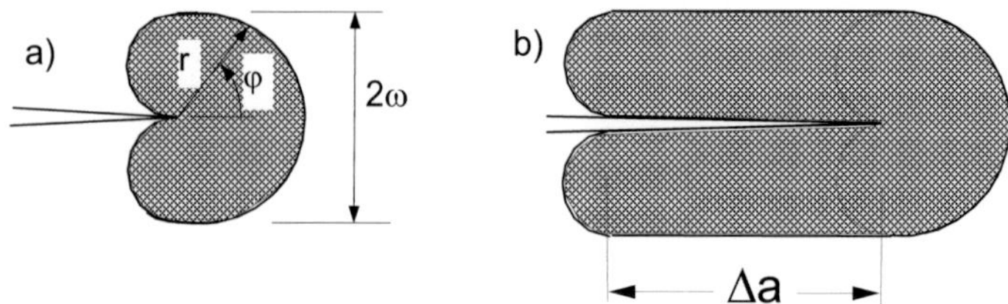

Fig. H1.1 a) Phase transformation zone ahead of a crack tip, b) zone after crack extension.

McMeeking and Evans [H1.1] computed the crack resistance curve under small-scale transformation conditions assuming weak transformations. This means that the singular stresses caused by the stress intensity factor only are considered, whereas the stresses caused by the phase transformations were neglected. From the analysis in [H1.1], the surface tractions result in a shielding stress intensity factor K_{sh}

$$K_{\mathrm{sh}} = p\oint_{\Gamma} \mathbf{n} \cdot \mathbf{h} \mathrm{d}S \qquad\qquad \text{(H1.1.5)}$$

where Γ is the contour line of the transformation zone and $\mathrm{d}S$ is a line length increment. The vector $\mathbf{h}$ represents the weight function $\mathbf{h} = (h_y, h_x)^{\mathrm{T}}$ with the components h_y and h_x. In the special case of a pure dilatational transformation, the surface tractions are given by the normal pressure p defined by

$$p = \frac{\varepsilon^{\mathrm{T}} fE}{3(1-2v)}, \qquad\qquad \text{(H1.1.6)}$$

where ε^T is the volumetric phase transformation strain, f the volume fracture of transformed material, and ν Poisson's ratio.

In Fig. H1.2, the shielding (residual) stress intensity factor for the case $T=0$, denoted as $K_{sh,0}$, is plotted for the phase transformation zone shown in Fig. H1.1b. The shielding stress intensity factor tends asymptotically to a value of 0.22.

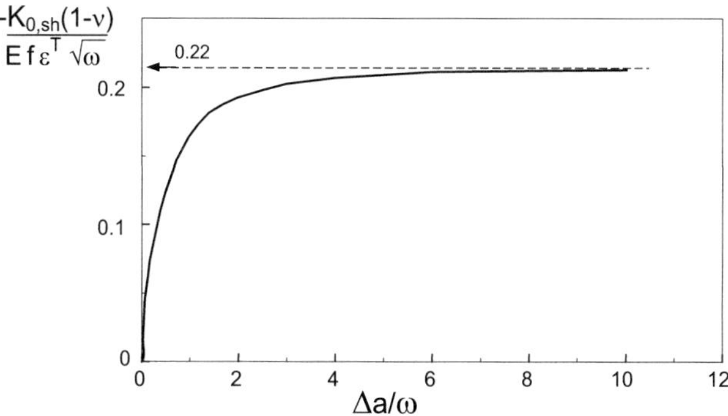

Fig. H1.2 Normalised shielding stress intensity factor $K_{0,sh}$ in the absence of a T-term computed with the method proposed by McMeeking and Evans [H1.1].

Because of (H1.1.4), it can be concluded that the shielding stress intensity factor K_{sh} must be proportional to the square root of the zone height

$$K_{sh} \propto \sqrt{\omega} \tag{H1.1.7}$$

with the factor of proportionality depending on the zone length Δa (curve in Fig. H1.2), the elastic constants E and ν, and the transformation strain ε^T.

H1.2 Micro-cracking zones

In poly-crystalline materials, e.g. ceramics, the high stresses ahead of a crack result in fracture of favourably oriented grain boundaries (Fig. H1.3). This micro-cracking at grain boundaries is caused by internal and superimposed externally applied stresses. The internal stresses are a consequence of thermal expansion mismatch in differently oriented grains. Different stress criteria for micro cracking were used in literature.

A critical value of the first invariant of the stress tensor (hydrostatic stress) was proposed by Evans and Faber [H1.5], an effective stress suggested in the study of Charalambides and McMeeking [H1.6].

In [H1.5], the shape and size of the micro-crack zone is assumed to be governed by the condition of a critical value of the hydrostatic stress being responsible for cracking, i.e.

$\sigma_{hyd}=\sigma_{hyd,cr}$. After micro cracking, the thermal mismatch stresses between the grains are released yielding an inelastic volume strain. Consequently, the same zone shape and the same type of the R-curve results.

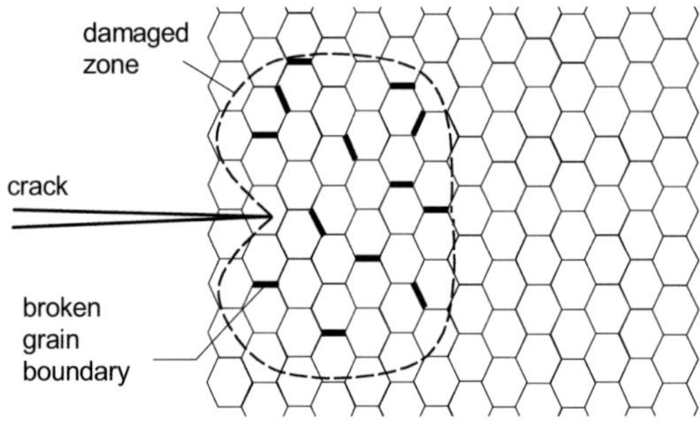

Fig. H1.3 Broken grain boundaries in a region ahead of a crack tip defining the micro-cracking zone.

References H1

H1.1 McMeeking, R.M., Evans, A.G. Mechanics of transformation-toughening in brittle materials, J. Am. Ceram. Soc. **65**(1982), 242–246.

H1.2 Budiansky, B., Hutchinson, J.W., Lambropoulos, J.C., Continuum theory of dilatant transformation toughening in ceramics, Int. J. Solids Struct. **19**(1983), 337–355.

H1.3 Giannakopoulos, A.E., Olsson, M., Influence of the non-singular stress terms on small-scale supercritical transformation toughness, J. Am. Ceram. Soc. **75**, pp. 2761-2764 (1992).

H1.4 Stam, G.T.M., van der Giessen, E., Meijers, P., Effect of transformation-induced shear strains on crack growth in zirconia-containing ceramics, Int. J. Solids Struct. **31**(1994), 1923–1948.

H1.5 Evans, A.G., Faber, K.T., Crack-growth resistance of microcracking brittle materials, J. Am. Ceram. Soc. **67**(1984), 255-260.

H1.6 Charalambides, P.G., McMeeking, R.M., Near-tip mechanics of stress-induced microcracking in brittle materials, J. Am. Ceram. Soc. **71**(1988), 465-472.

H2 R-curve effect for glass by ion exchange

In the preceding Sections an R-curve behaviour was defined by the existence of a shielding stress intensity factor (due to crack-face bridging or phase transformation around a crack tip) which reduced the applied load and was dependent on the amount of crack propagation Δa. This point of view allows also treating shielding effects in glasses by R-curve behaviour. In this context, it has to be mentioned that glasses are considered as the standard materials, which should not show any R-curve effect because crack-face interactions and phase transformations are not present.

The reason for an R-curve effect is the occurrence of an ion exchange layer at glass surfaces in contact to water or humid air. Several indications for an ion exchange are known. The measurement of hydrogen profiles is addressed here as the possibly most evidential example.

H2.1 Ion exchange layers in soda-lime glass

Soda-lime-silicate glasses are very corrosion resistant at room temperature, but do react with water to form a thin hydration layer on the glass surface. Surface hydration consists of the interdiffusion of either hydrogen ions (H^+), or hydronium ions (H_3O^+), with the Na^+ ions in the glass. The H^+/Na^+ exchange results in a tensile stress in the hydration layer, because the H^+ ion is smaller than the Na^+ ion. By contrast, the H_3O^+/Na^+ exchange leads to a compressive stress, because H_3O^+ is larger than Na^+. Measurements of the hydrogen and soda profiles [H2.1] revealed a ratio between the hydrogen concentration in the surface of the hydrated glass and the sodium concentration in the unhydrated glass of about 3. This three-to-one replacement of Na^+ by H^+ suggests that Na^+ is replaced by H_3O^+ resulting in a volume expansion.

The ion exchange behaviour of the glass is shown in Figure H2.1. Two hydrogen profiles measured in [H2.1] are shown in Fig. H2.1a in a normalized plot with the concentration c_0 at the surface and the depth y nomalized on that depth b at which a reduction of 50% was observed.

Figure H2.1b shows the thickness b of the ion exchange layer as a function of time t. The $b \propto \sqrt{t}$ dependency for the *nearly step-shaped* profiles suggests a description by

$$b = 2\sqrt{Dt} \qquad\qquad (\text{H2.1.1})$$

defining the temperature dependent diffusivity $D(\Theta)$. Figure H2.1c shows the layer thickness after 22 h at several temperatures in an Arrhenius plot. This result indicates that the growth of the layer thickness is strongly enhanced with increasing temperature Θ.

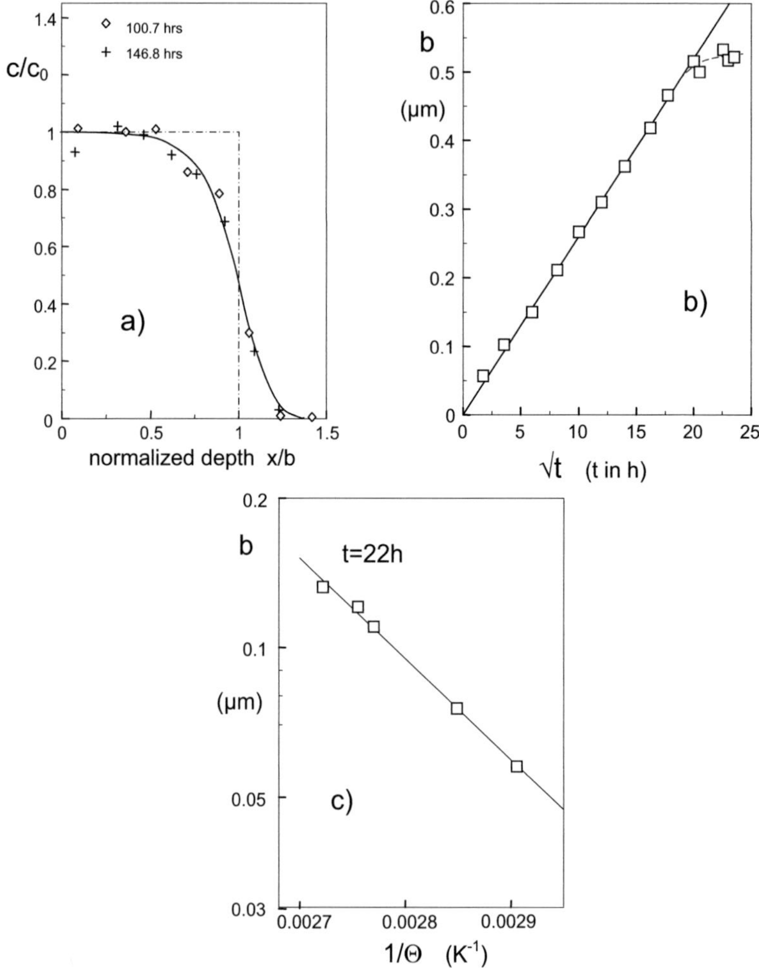

Fig. H2.1 Thickness of ion exchange layers for soda-lime glass in water according to Lanford et al.[H2.1]; a) hydrogen profile (water at 90°C), b) influence of time, c) influence of temperature.

Additional indications are obvious for instance from measurements of deformations [H2.2] and strengths [H2.3].

H2.2 Stress intensity factors for ion exchange layers

H2.2.1 Computation procedure

The ion exchange layers give rise for an intrinsic shielding stress intensity factor that shields the crack tip partially from an externally applied load. First attempts for its computation were carried out by Bunker and Michalske [H2.4] and Michalske *et al.* [H2.5]. These authors used fracture mechanics formalism, originally developed to interpret the fracture toughness of ZrO_2 [H2.6].

A surface layer is considered in which a constant volumetric strain ε is generated by ion exchange (Fig. H2.2). McMeeking and Evans [H2.6] developed the procedure for the computation of mode-I stress intensity factors (see Section H1).

The related shielding stress intensity factor K_{sh} is given by the contour integral

$$K_{sh} = p\int_{\Gamma} \mathbf{n} \cdot \mathbf{h} dS \qquad\qquad (H2.2.1)$$

with the normal vector $\mathbf{n}$ on the zone contour and the normal pressure p defined by

$$p = \frac{\varepsilon E}{3(1-2v)}, \qquad\qquad (H2.2.2)$$

where E is Young's modulus, v Poisson's ratio, and ε the volumetric strain. Γ is the contour line of the zone and dS is a line length increment. The vector $\mathbf{h}$ represents the weight function $\mathbf{h}_I = (h_{I,y}, h_{I,x})^T$ with the components $h_{I,y}$ and $h_{I,x}$

$$h_{I,x} = \frac{1}{\sqrt{8\pi r}(1-v)}[2v-1+\sin(\theta/2)\sin(3\theta/2)]\cos(\theta/2) \qquad (H2.2.3a)$$

$$h_{I,y} = \frac{1}{\sqrt{8\pi r}(1-v)}[2-2v-\cos(\theta/2)\cos(3\theta/2)]\sin(\theta/2). \qquad (H2.2.3b)$$

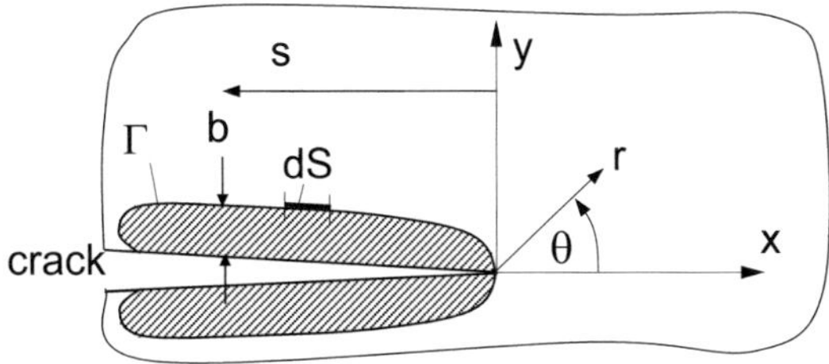

Fig. H2.2 Crack with a crack face zone undergoing volumetric strains.

In [H2.7], the two limit cases for the profile of the ion exchange layer were considered, one for a moving crack, the other for an arrested crack. In the related shielding stress intensity factor analysis, the case of ion exchange with a stress independent diffusion constant D was used. In this context it should be mentioned that diffusion of water in silica is strongly stress enhanced (Section H3).

In the following considerations, the case of the moving crack may be repeated here in terms of an R-curve description.

H2.2.2 Stress intensity factors for growing cracks

A crack propagating with a constant rate $v=\Delta a/t$ over a sufficiently long distance shows a layer thickness profile near the tip according to Fig. H2.3a, represented by

$$b = b_{max}\sqrt{\frac{s}{\Delta a}}, \quad b_{max} = 2\sqrt{\frac{D\Delta a}{v}} \qquad (\text{H2.2.4})$$

with the distance s from the crack tip. In this case the shielding mode-I stress intensity factor is

$$K_{sh} = -\frac{\varepsilon E}{1-v}\frac{8\sqrt{D/v}}{3\sqrt{15}}\arctan\left[\frac{\frac{2}{3}\sqrt{\frac{5}{3}}\sqrt{v\,\Delta a/D}}{1+\frac{5}{27}(1+2\sqrt{v\,\Delta a/D})}\right] \qquad (\text{H2.2.5})$$

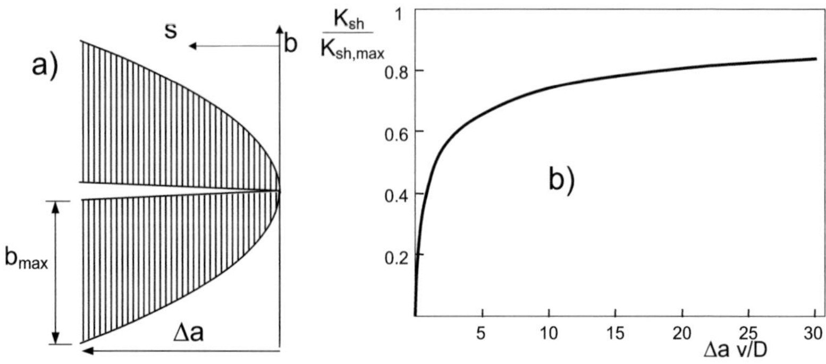

Fig. H2.3 a) Square-root shaped volume strain regions as expected for cracks in glass growing at constant rate, b) shielding stress intensity factor K_{sh} as a function of crack extension (normalized representation).

The maximum shielding term for $\Delta a\rightarrow\infty$ is

$$K_{sh,max} \cong -\frac{\varepsilon E}{1-\nu}\,\frac{8\sqrt{D/v}}{3\sqrt{15}}\,\arctan\!\left[\sqrt{\frac{27}{5}}\,\right] \qquad\text{(H2.2.6)}$$

and the normalised representation then reads

$$\frac{K_{sh}}{K_{sh,max}} \cong 0.859\,\arctan\!\left[\frac{\frac{2}{3}\sqrt{\frac{5}{3}}\sqrt{v\,\Delta a/D}}{1+\frac{5}{27}(1+2\sqrt{v\,\Delta a/D})}\right] \qquad\text{(H2.2.7)}$$

This dependency is illustrated in Fig. H2.3b showing the typical R-curve behaviour as obtained for ceramics with crack-face bridging and phase-transformation toughening.

Next, a crack of initial length a_0 is considered as shown in Fig. H2.4a. This crack may be loaded by an external load. At time $t=0$, liquid water or humid air may appear along the crack and subcritical crack growth will start abruptly. The zone height b results from (H2.1.1) as

$$b(s)=\begin{cases} 2\sqrt{D\,s/v} & for \quad 0\le s<\Delta a \\ 2\sqrt{D\,\Delta a/v}\ \,(=b_{max}) & for \quad s\ge \Delta a \end{cases} \qquad\text{(H2.2.8)}$$

This profile is illustrated in Fig. H2.4a. The related shielding stress intensity factor is shown in Fig. H2.4b. The dashed curve again shows the result for the parabolic profile, Fig. H2.3. The differences are small for $\Delta a\,v/D>1$. Consequently, eq.(H2.2.7) may be used for an approximate description. Figure H2.4c shows the shielding stress intensity factor in the more convenient form as a function of $\Delta a/b$.

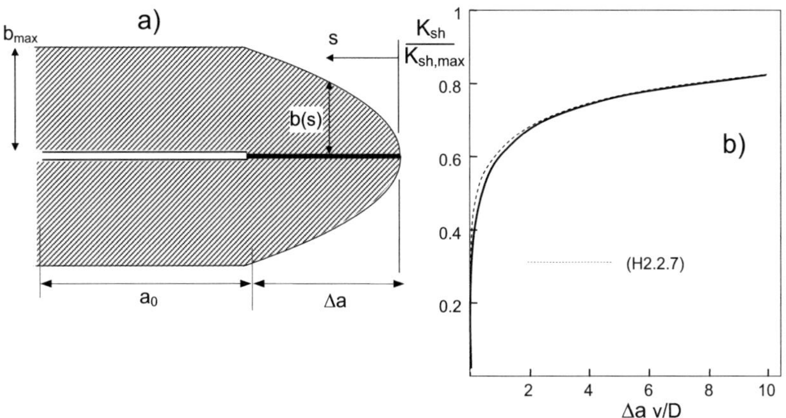

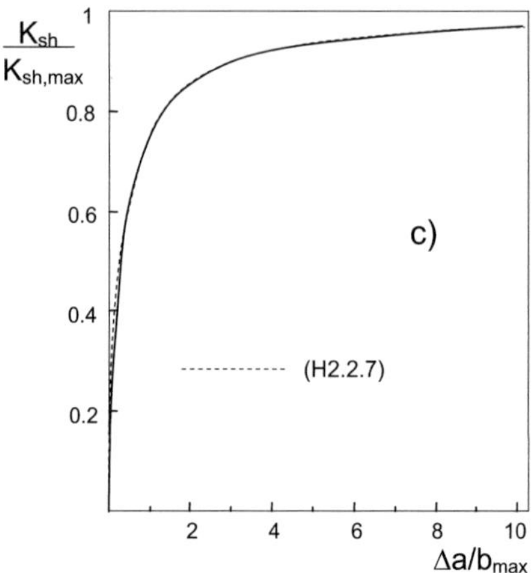

Fig. H2.4 Development of an ion exchange layer for subcritical crack extension at a constant crack growth rate, a) zone shape, b) normalized R-curve (solid curve), dashed curve: R-curve from Fig. H2.3b), c) R-curve in the form $K_{sh}=f(\Delta a/b_{max})$.

H2.2.3 Shielding stress intensity factor for an arrested crack

Whereas the growing crack shows a parabolic layer thickness profile close to the crack tip, an arrested crack ($v\rightarrow 0$) will exhibit a zone as shown by the two limit cases in Fig. H2.5.

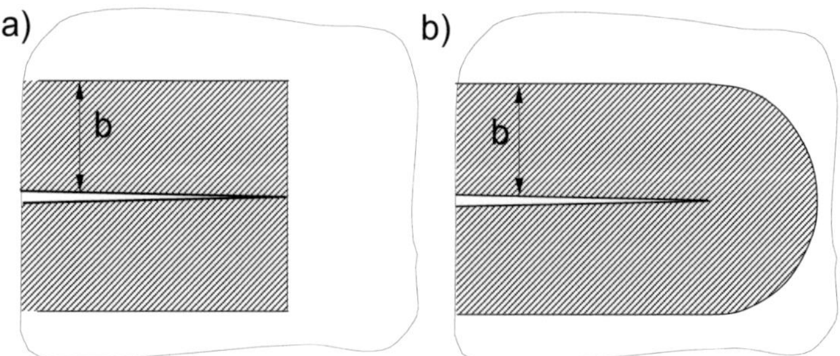

Fig. H2.5 Limit cases of ion exchange layer profiles for an arrested crack; a) diffusion only in height direction, b) diffusion in height and crack-plane directions.

The shielding stress intensity factor for the case of Fig. H2.5a can be described by

$$K_{sh} = -0.374\frac{\varepsilon E}{1-v}\sqrt{b} \overset{(H2.1.1)}{=} -0.374\frac{\varepsilon E}{1-v}\sqrt{2}(Dt)^{1/4} \qquad (H2.2.9a)$$

For thicker zones (i.e. longer time t) the diffusion <u>in</u> crack direction can no longer be neglected. It will result in a layer formed around the crack tip as illustrated by Fig. H2.5b. If the shape of the extended part is represented by a half-circle, the related shielding stress intensity factor is given by [H2.6, H2.7]

$$K_{sh} = -0.25 \frac{\varepsilon E}{1-\nu} \sqrt{b} \overset{(H2.1.1)}{=} -0.25 \frac{\varepsilon E}{1-\nu} \sqrt{2} (Dt)^{1/4} \qquad \text{(H2.2.9b)}$$

Apart from the coefficients (0.25 and 0.374), both zone shapes show the same result.

H2.3 Consequence on threshold of subcritical crack growth

As outlined in Section H2.2, the maximum shielding term $|\Delta K|_{sh,max}$ (see eq.(H2.2.6)) depends on the subcritical crack growth rate v according to

$$K_{sh,max} \propto \sqrt{\frac{1}{v}} \qquad \text{(H2.3.1)}$$

In a subcritical crack growth test, the two conditions must hold:

1) The total stress intensity factor K_{total} representing the stress singularity at the crack tip (often denoted as K_{tip}) is given by superposition of applied and shielding terms

$$K_{total} = K_{tip} = K_{appl} + K_{sh} \;, \quad K_{sh} < 0 \qquad \text{(H2.3.2)}$$

2) At identical subcritical crack growth rates v, the total stress intensity factor must be the same.

Subcritical crack growth can often be described by a power-law relation

$$v = A K_{tip}^{n} \qquad \text{(H2.3.3)}$$

with a high exponent $n>10$. Consequently, the maximum shielding stress intensity factor as a function of K_{tip}, eq.(H2.3.1) reads

$$K_{sh,max} \propto - K_{tip}^{-n/2} \qquad \text{(H2.3.4)}$$

i.e. the shielding term increases strongly with decreasing crack-tip stress intensity factor.

The applied stress intensity factor then results from (H2.3.2)

$$K_{appl} = K_{tip} + c_1 K_{tip}^{-n/2} \qquad \text{(H2.3.5)}$$

This dependency is plotted in Fig. H2.6a. The arrow represents the shielding stress intensity factor. The minimum value of the applied stress intensity factor $K_{appl,min}$ (indicated by the circle) results from the condition $dK_{appl}/dK_{tip}=0$

$$K_{appl,min} = c_1\left[\left(\frac{nc_1}{2}\right)^{\frac{n}{n-2}} + \left(\frac{2}{nc_1}\right)^{\frac{2}{n-2}}\right] \qquad (H2.3.6)$$

with the related K_{tip} of

$$K_{tip}(K_{appl,min}) = \left(\frac{2}{nc_1}\right)^{\frac{2}{n-2}} \qquad (H2.3.7)$$

Figure H2.6b shows the v-K_{tip}-curve according to (H2.3.3) as the dash-dotted straight line. A crack-growth test may be considered in which the externally applied load is continuously and slowly reduced in order ensure steady-state conditions with a sufficiently large crack increment Δa grown (arrow 1). For decreasing K_{appl} the v-K_{appl}-curve becomes steeper until $K_{appl,min}$ the point of infinite steepness is reached. The dashed curve part below this point cannot be reached because this would need an increase of K_{appl} again.

The minimum shielding stress intensity factor $K_{appl,min}$ does of course not imply that the related crack velocity $v=v(K_{appl,min})$ would be the lowest reachable velocity in the crack-growth test described before. $K_{appl,min}$ is the minimum shielding term that is reached only under conditions of constant subcritical crack growth rates v and sufficiently large crack increments Δa. A computation of the crack growth rate below the circle (along arrow 2) varying with time would be an excessive demand for the simple model.

A more extended analysis of this problem was given in [H2.7] on the basis of the limit cases of a moving crack with $K_{sh}\propto 1/\sqrt{v}$ as considered in Section H2.2.2 and an arrested crack with the shielding term depending on time $K_{sh}\propto t^{1/4}$.

Figure H2.7a shows the two shielding stress intensity factor solutions for a crack growing at a constant rate (horizontal dashed line) and for an arrested crack (dash-dotted line). These two solutions intersect for a certain time t_0.

In [H2.7] an approximate evaluation of shielding stress intensity factors under varying time and crack velocity was performed by interpolation of the two limit cases valid for very short and very long times. The result was

$$K_{sh}(t) = \frac{K_{sh}(t \to \infty)}{(1+t_0/t)^{1/4}} \qquad (H2.3.8)$$

as shown in Fig. H2.7a by the continuous curve. The shielding stress intensity factor as a function of time is represented in Fig. H2.7b.

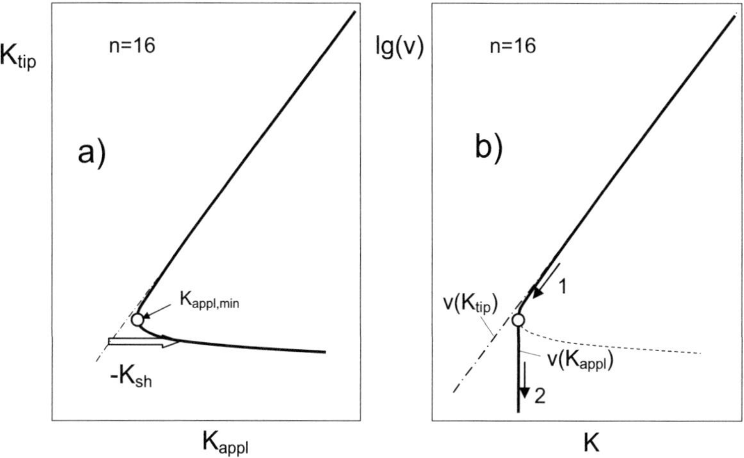

Fig. H2.6 a) Maximum shielding for crack extension at constant crack velocity, b) crack growth curves $v(K_{tip})$ (dash-dotted line) and $v(K_{appl})$ (solid curve).

Interpolation according to eqs.(H2.3.8) then yields the applied stress intensity factor

$$K_{appl} = K_{tip} + 0.4\frac{2\sqrt{D}}{\sqrt{A}K_{tip}^{n/2}}\frac{\varepsilon E}{1-\nu}\frac{1}{\left(1+\dfrac{26D}{t}\left(\dfrac{1}{AK_{tip}^{n}}\right)^{2}\right)^{1/4}} \qquad (H2.3.9)$$

Figure H2.8a represents the total stress intensity factor K_{tip} as a function of K_{appl} according to eq.(H2.3.9) for different times. If the applied load is reduced below the threshold condition (circle), then K_{tip} decreases with time because of the growth of the exchange layer thickness on the sides of the crack.

The ordinate in Fig. H2.8a can be converted into a crack velocity by the use of eq.(H2.3.3). Crack velocity, v, is then obtained as a function of applied stress intensity factor K_{appl} (Fig. H2.8b). At high values of $K_{,appl}$ a single-valued curve is obtained relating v and K_{appl}. When the load is reduced below the threshold value, the single curve separates into a set of curves that depend on time. For a fixed value of K_{appl} the crack velocity continuously decreases with time. The crack asymptotically stops growing; the condition of crack arrest is reached.

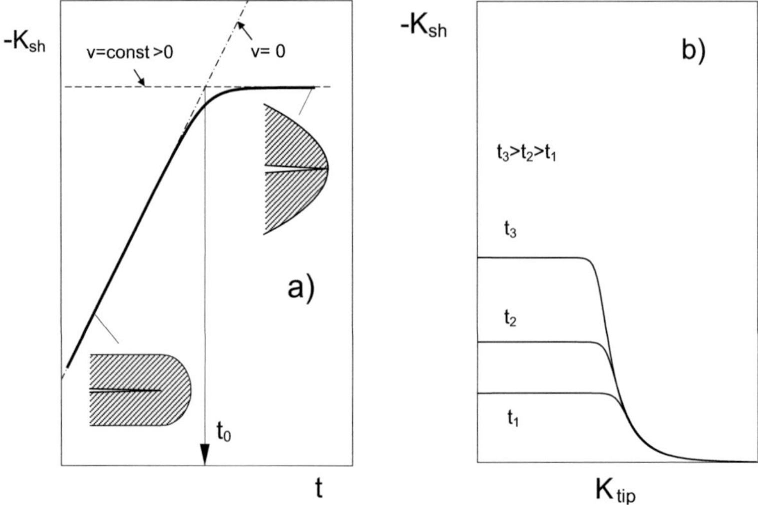

Fig. H2.7 a) Computation of the intrinsic stress intensity factor K_{sh} for cracks growing with different rates (straight lines: $K_{sh}=\propto 1/\sqrt{v}$ and $K_{sh}\propto t^{1/4}$), curve: interpolation according to eq.(H2.3.8)).

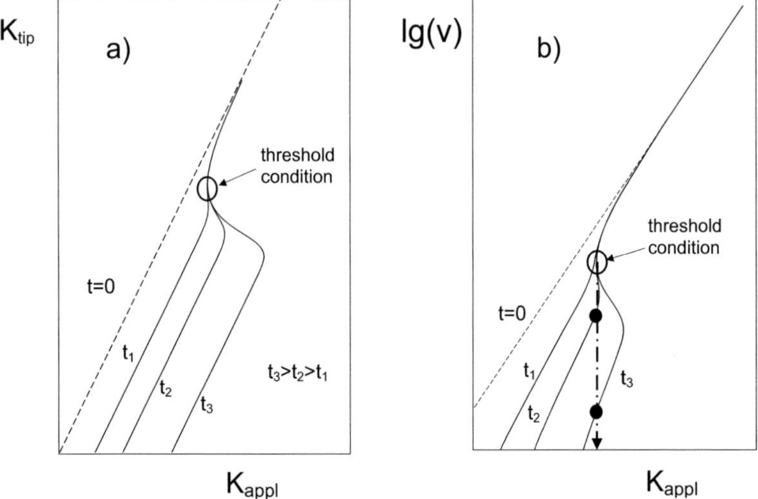

Fig. H2.8 a) Applied stress intensity factor as a function of the total stress intensity factor K_{tip} and time t, b) decrease of the crack growth rate with time.

H2.4 Effect of Na$^+$-H$^+$ ion exchange

So far, we have considered ion exchange alkali$^+$↔H$_3$O$^+$. For other glass compositions and environments in which the exchange alkali$^+$↔H$^+$ prevails, the same relations derived before hold with a changed sign in the shielding stress intensity factor (K_{sh}>0) because the H$^+$ ion needs a smaller volume than the alkali ion. In this case, ion exchange layers must enhance subcritical crack growth. The expected K_{tip}-K_{appl} and v-K_{appl}-curves are schematically shown in Fig. H2.14. Instead of the threshold behaviour, a plateau in the v-K_{appl}-curve has to be expected. Such behaviour has been observed by Simmons and Freiman [H2.8] and Gehrke et al. [H2.9, H2.10, H2.11].

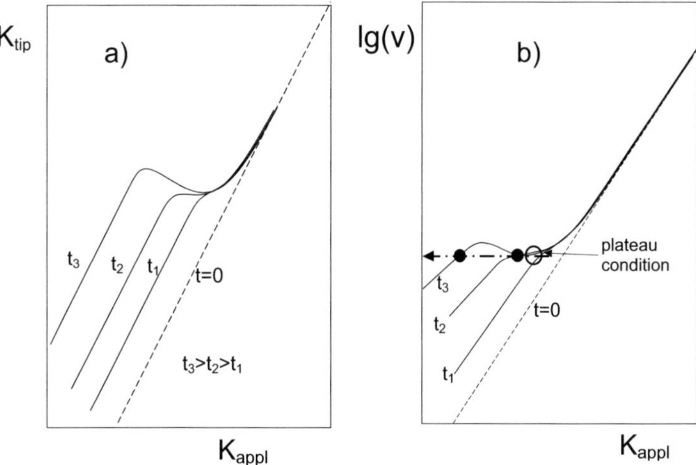

Fig. H2.9 a) Applied stress intensity factor as a function of the total stress intensity factor K_{tip} and time t, b) change of the crack growth rate with time.

Remark:

The shielding term by ion exchange layers is in principle very similar to the shielding effects mentioned in Section H1. However, one has to have in mind that the R-curve effect as an increase of the shielding term is an effect on the length scale of a few nm. Saturation is already reached after crack extensions of less than 1μm. From this point of view, glass may be considered as a material without an R-curve effect in the usual sense.

References H2

H2.1 Lanford, W.A., Davis, K., Lamarche, P., Laursen, T., Groleau, R., Doremus, R.H., Hydration of soda-lime glass, J. Non-Cryst. Sol. **33** 249-266 (1979).

H2.2 Fett, T., Guin, J.P., Wiederhorn, S.M., Stresses in ion-exchange layers of soda-lime-silicate glass, Fatigue and Fracture of Engineering Materials and Structures, **28**(2005),507-14.

H2.3 Fett, T., Guin, J.P., Wiederhorn, S.M., Estimation of ion exchange layers for soda-lime-silicate glass from curvature measurements, J. Mater. Sci. **41**(2006), 5006–50

H2.4 Bunker, B.C., Michalske, T.A., pp. 391-411 in "Effect of surface corrosion on glass fracture", *Fracture Mechanics of Ceramics, Vol. 8*, Plenum Press, New York (1986).

H2.5 Michalske, T.A., Bunker, B.C., Keefer, K.D., Mechanical properties and adhesion of hydrated glass surface layers, J. Non-Cryst. Sol. **120**(1990), 126-137.

H2.6 McMeeking, R.M., Evans, A.G., Mechanics of Transformation-Toughening in Brittle Materials, J. Am. Ceram. Soc. **65**[5] 242-246 (1982).

H2.7 Fett, T., Guin, J.P., Wiederhorn, S.M., Interpretation of effects at the static fatigue limit of soda-lime-silicate glass, Engng. Fract. Mech. **72**(2005), 2774-279.

H2.8 Simmons, C.J., Freiman, S.W., Effect of corrosion processes on subcritical crack-growth in glass, J. Am. Ceram. Soc. 64(1981), 683-686.

H2.9 Gehrke, E., Ullner, C., Hähnert, M., Fatigue Limit and Crack Arrest in Alkali-Containing Silicate Glasses, J. Mater. Sci. **26**(1991), 5445-5455.

H2.10 Gehrke, E., Ullner, C., Application of fractography of glass to the study of environmental effects on crack growth," Advances in Ceramics, Vol. 22: Fractography of Glasses and Ceramics, 1988, 77-84.

H2.11 Gehrke, E., Ullner, C., Hähnert, M., Effect of corrosive media on crack growth of model glasses and commercial silicate glasses, Glastech. Ber. **63**(1990), 255-265.

H3 Crack-tip shielding in silica

In the preceding Section, an R-curve effect in glass was discussed based on the ion exchange in glass with alkali content. The main reasons for the effect were diffusion and different size of exchanged ions. In the following considerations, a possible effect of diffusion of water molecules into the glass structure will be discussed. The diffusion of water is rather strong and there is some evidence for a related volume expansion. This combination could also be responsible for R-curve behaviour. Whereas the diffusion coefficients for many glasses are known, there is actually a lack in knowledge about the volume change in glass due to water penetration. In this respect, the following sections are somewhat hypothetical. Since the most evidence for a volume change seems to be present for silica, this material will be taken into account in the following sections.

H3.1 Water diffusion and volume expansion in silica

H3.1.1 Diffusion

Water can diffuse into silica glass surfaces. This effect is temperature dependent and described by the diffusion coefficient D according to

$$D_w = A_0 \exp(-Q/R\Theta) \tag{H3.1.1}$$

with the activation energy Q, the absolute temperature Θ, and the gas constant R. As reported in reference [H3.1]: $Q=71.2\pm1$ kJ/mol, $\log_{10} A_0 = -7.78\pm0.06$ (A_0 in m^2/s) for *molecular* water and $Q=72.3$ kJ/mol, $\log_{10} A_0 = -8.12$ for the *effective* diffusivity (temperature range 0°C to 200°C).

The water concentration profile as a function of depth y and time t, $C(y,t)$, is given by

$$C(y,t) = C_0 \, \mathrm{erfc}\!\left(\frac{y}{2\sqrt{Dt}}\right) \tag{H3.1.2}$$

where C_0 is the surface concentration and erfc represents the complementary error function. In this equation it is implicitly assumed that stresses by swelling are negligible.

In the presence of tensile stresses the diffusion is enhanced as measured by Nogami and Tomozawa [H3.2]. The diffusion coefficient as a function of the hydrostatic stress σ_h could be written

$$D = D_0 \exp\left(\frac{\Delta V_w \sigma_h}{R\Theta}\right), \quad D_0 = D_w(\sigma_h = 0) \tag{H3.1.3}$$

with the activation volume ΔV and the diffusivity at zero hydraulic pressure D_0. For room temperature, $\Delta V \approx 10\text{-}30$ cm^3/mol [H3.3]. This is very large compared to the activation volume ΔV for alkali ions in silicate glass [H3.2]

H3.1.2 Swelling

In literature, there is clear evidence for swelling measured directly via dimension change and indirectly via stress generation.

The swelling behaviour has been determied directly by length [H3.4] and curvature measurements on bars [H3.5] and X-ray diffraction strain measurements [H3.6].

Gorbacheva and Zaoints [H3.4] soaked fused quartz prisms (25×5×5 mm) in water at 80°C for 20 months. Then they measured the length change and found an increase of 0.17%. Since only a thin surface layer could have been generated during the water storage, the suppressed linear strain in the surface must be clearly larger.

In a birefrigerence evaluation [H3.7] the occurrence of a swelling effect could be proved indirectly.

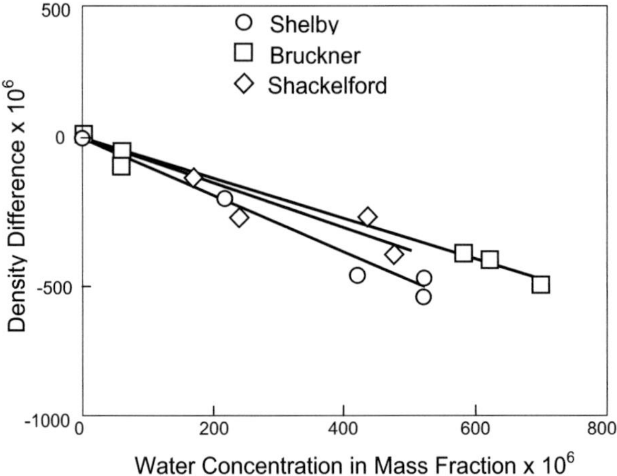

Fig. H3.1 Effect of water concentration on density of vitreous silica, results by Shelby [H3.8], Bruckner [H3.9, H3.10], and Shackelford [H3.11].

Quantitative swelling data from density measurements on vitreous silica were reported by Shelby [H3.8]. Figure H3.1 shows results by Shelby [H3.8], Bruckner [H3.9,

H3.10] and Shackelford [H3.11]. The observed linear decrease of density ρ as a function of the water concentration C_w can be expressed by a simple linear dependency

$$\frac{\Delta\rho}{\rho} = -\chi\,C_w \quad , \quad \chi = \begin{cases} 0.96 & \text{Shelby} \\ 0.84 & \text{Shackelford} \\ 0.71 & \text{Bruckner} \end{cases} \tag{H3.1.4}$$

From the definition of the density as the quotient of mass m and volume V, $\rho=m/V$, it follows by taking logarithmic derivations

$$V = \frac{m}{\rho} \;\Rightarrow\; \varepsilon_v = \frac{\Delta V}{V} = \frac{\Delta m}{m} - \frac{\Delta\rho}{\rho} \tag{H3.1.5}$$

Since only the water content m_w can change, the mass change is

$$\Delta m = m_w = C_w m \tag{H3.1.6}$$

and the volume swelling strain ε_v simply results by introducing (H3.1.4) and (H3.1.6) in (H3.1.5) as

$$\varepsilon_v = (1+\chi)\,C_w \tag{H3.1.7}$$

H3.1.3 Stresses due to swelling

At a free surface, the stress state is plane stress and, consequently, also stresses caused by swelling are equi-biaxial ($\sigma_y=0$)

$$\sigma_x = \sigma_z = -\frac{\varepsilon_v(y,t)E}{3(1-\nu)} \tag{H3.1.8}$$

where E is Young's modulus and ν is Poisson's ratio. Consequently, the hydrostatic stress reads

$$\sigma_h = \tfrac{1}{3}(\sigma_x + \sigma_z) = -\frac{2\,\varepsilon_v(y,t)E}{9(1-\nu)} \tag{H3.1.9}$$

Since the diffusion coefficient depends on the hydrostatic stress component according to eq.(H3.1.3), the hydrostatic stress due to swelling again must affect the concentration profile. Equation (H3.1.2) is then no longer correct with the consequence that the diffusion differential equation has to be solved numerically.

H3.2 Approximate 1-dim treatment of the crack-tip swelling zones

H3.2.1 Swelling zone for unloaded cracks

In the absence of a crack-tip stress field (i.e. for an unloaded crack) water diffusion yields a diffusion zone as represented in Fig. H3.2a.

Water entering the glass through the plane of the crack is basically a one-dimensional diffusion problem, with the solution given by eq.(H3.1.2). Near the crack tip and ahead of the crack a much more complicated two-dimensional non-axial solution to the diffusion equations is required. For reasons of clearness two assumtions are made:

1. The diffusion zone along the crack of length a (Fig. H3.2a) is assumed to be terminated by a circular part around the crack tip.

2. The continuously varying swelling strain $\varepsilon(y)$ is replaced by a step-shaped dependency with a change at the characteristic contour height $b \cong y_{1/2}$ at which the water concentration and consequently the swelling strain are reduced to 50 % of the surface values.

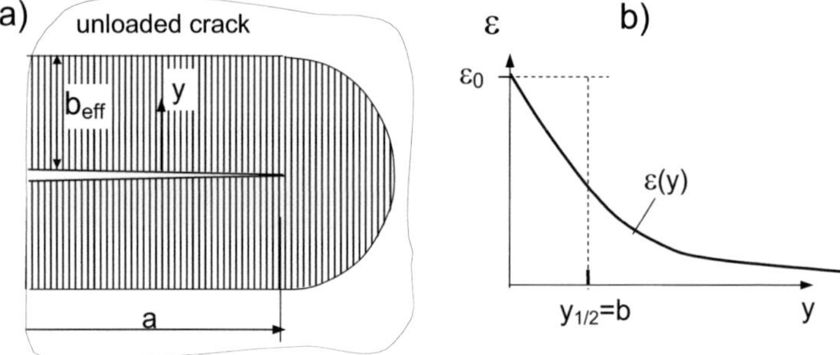

Fig. H3.2 a) Diffusion zone in the absence of an external load, b) swelling profile under assumption of one-dimensional; definition of an effective zone thickness b by a reduction of water content of 50%.

The volume expansion at crack surfaces must result in an intrinsic shielding stress intensity factor K_{sh}. In the case of an unloaded crack, the hydrostatic stress term in eq.(H3.1.3) disappears. Because of the Arrhenius dependence of the diffusivity on temperature, eq.(H3.1.1), the diffusion process becomes more important as the temperature is increased, even though no stress-enhancement takes place. The shielding stress intensity factor, K_{sh}, after a time, t, of water contact is given by

$$K_{sh} = -0.25 \frac{\varepsilon_0 E}{1-\nu} \sqrt{b} \qquad (H3.2.1)$$

with an effective water layer thickness, b, of

$$b \cong \sqrt{D_0 \cdot t} \qquad (\text{H3.2.2})$$

and the volumetric strain ε_0 at the surface, $\varepsilon_0 = \varepsilon_v(y=0)$. Finally, it holds that

$$K_{sh} \cong -0.25 \frac{\varepsilon_0 E}{1-v}[D_0\, t]^{1/4} \qquad (\text{H3.2.3})$$

H3.2.2 Swelling zone for a non-propagating crack under load

The hydrostatic stress is given by the trace of the stress tensor

$$\sigma_h = \tfrac{1}{3}(\sigma_{rr} + \sigma_{\varphi\varphi} + \sigma_{zz}) \qquad (\text{H3.2.4})$$

Since very high stresses occur in the vicinity of a crack tip, a water containing zone must rapidly extend during crack growth.

If K_{tip} denotes the total stress intensity factor, the singular near-tip stress field reads

$$\sigma_{ii} = \frac{K_{tip}}{\sqrt{2\pi r}} g_{ii}(\varphi) \qquad (\text{H3.2.5})$$

where r is the crack-tip distance. In eq.(H3.2.5) g_{ii} are well known geometric functions depending on the polar angle φ, Fig. H3.3a.

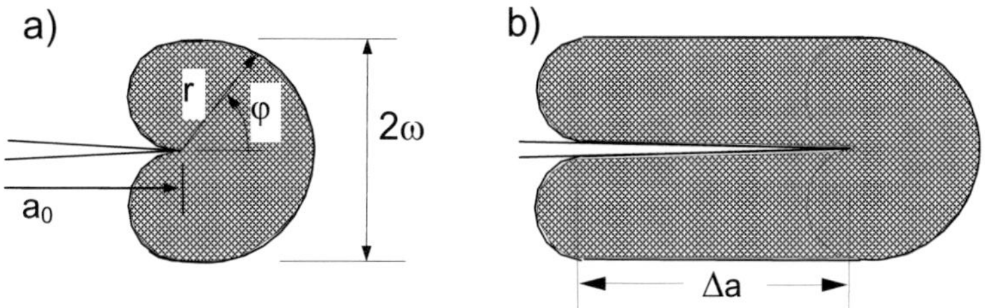

Fig. H3.3 a) Swelling zone at the tip of an arrested crack under mechanical loading, caused by stress-enhanced diffusion, b) zone for a crack grown by Δa.

The hydrostatic stress under plane strain conditions is

$$\sigma_{h,tip}(K_{tip}) = \tfrac{2}{3}(1+v)\frac{K_{tip}}{\sqrt{2\pi r}}\cos(\varphi/2) \qquad (\text{H3.2.6})$$

In the crack tip region the diffusion coefficient is strongly affected by the singular stresses. For a simple 1-dimensional analysis [H3.3] in which only the stresses caused by the externally applied load were taken into account, the water concentration profile could be derived by introducing eq.(H3.2.4) into (H3.1.3) and using the result in eq.(H3.1.2). The radial distance from the tip at which the water concentration and the swelling strain are ½ of their surface values is for a loaded but arrested crack, Fig. H3.3a,

$$ r_{1/2} = \frac{\kappa^2 \cos^2(\varphi/2)}{4\left(PLog\left[\frac{\kappa \cos(\varphi/2)}{2 \cdot [D_0\, t]^{1/4}} \right] \right)^2} \quad , \quad \kappa = \frac{(1+v)}{3} \frac{K_{tip}}{\sqrt{2\pi}} \frac{\Delta V_w}{R\Theta} \tag{H3.2.7} $$

where the "PLog" stands for the Lambert W function or *product log function*, i.e. the solution $W=PLog(z)$ of the equation $z=W \exp(W)$.

The somewhat unusual expression (H3.2.7) in terms of the product logarithm is a consequence of perturbation theory adopted to the effect of stresses on diffusivity. For its application the effect of the hydrostatic stresses was dealt as a disturbance of the diffusion problem. Perturbation theory suggests to solve the problem for the undisturbed diffusivity and to insert then the disturbance parameter in this solution.

The compressive stresses caused by swelling can reach values in the order of GPa. The crack-tip swelling zone, eq.(H3.2.7), was computed in [H3.3] for the case of "weak swelling" conditions, neglecting any perturbation of the crack-tip stress field by the swelling strains similar to the case of "weak phase transformation" in zirconia ceramics. The latter assumes that the intrinsic stress field caused by a phase transformation does not affect the stress field triggering the transformation [H3.12, H3.13, H3.14].

In the case of a crack grown by a crack increment of Δa the diffusion zone extends as plotted in Fig. H3.3b.

H3.3 Two-dimensional solution of the diffusion equation

In a refined procedure, these high stresses were included in the full solution of the crack-tip diffusion problem [H3.15]. For variable diffusivity the equation for diffusion in the r-φ-plane has the following form:

$$ \frac{\partial C_w}{\partial t} = \nabla(D\nabla C_w) = \nabla D \nabla C_w + D\Delta C_w \tag{H3.3.1} $$

In the bulk material plane strain conditions prevail and any gradient with respect to the crack front disappears (i.e. $\partial/\partial z=0$). Equation (H3.3.1) can then be written in polar coordinates

$$\frac{\partial C_w}{\partial t} = \frac{1}{r}\frac{\partial}{\partial r}\left(r\,D\,\frac{\partial C_w}{\partial r}\right) + \frac{1}{r^2}\frac{\partial D}{\partial \varphi}\frac{\partial C_w}{\partial \varphi} + \frac{D}{r^2}\frac{\partial^2 C_w}{\partial \varphi^2} \tag{H3.3.2}$$

The diffusivity is given by

$$D = D_0\,\exp\left[(\sigma_{h,tip} + \sigma_{h,swell})\frac{\Delta V_w}{R\Theta}\right] \tag{H3.3.3}$$

with the hydrostatic swelling stress

$$\sigma_{h,swell} \cong -\gamma\,\frac{\varepsilon_v(r,t)E}{(1-\nu)} \tag{H3.3.4}$$

In order to determine the coefficient γ for a cordial-shaped zone (continuous curve in Fig. H3.4) undergoing a constant volume strain $\varepsilon_v = \varepsilon_0 = $ constant, the hydrostatic swelling stress $\sigma_{h,swell}$ was evaluated at a large number of nodes of an FE-mesh under plane strain conditions. It has to be mentioned that such a zone does not create a shielding stress intensity factor as had been shown by McMeeking and Evans [H3.12].

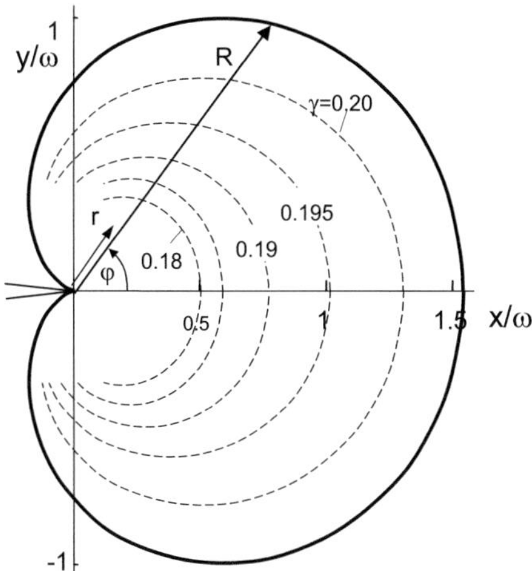

Fig. H3.4 Contour plots for constant hydrostatic swelling strains $\sigma_{h,swell}$ expressed by the coefficient γ of eq.(H3.3.4).

The hydrostatic swelling stresses expressed by the coefficient γ defined by eq.(H3.3.4) are shown by the dashed contours. The result under plane strain was obtained by an average over the zone radius R

$$\bar{\gamma} = \frac{1-\nu}{\varepsilon_0 E} \frac{1}{R} \int_0^R \sigma_{h,swell} \, dr \approx 0.19 \qquad (H3.3.5)$$

or as an average over the zone cross-section A

$$\bar{\gamma} = \frac{1-\nu}{\varepsilon_0 E} \frac{1}{A} \int_{(A)} \sigma_{h,swell} \, dA' \approx 0.195 \qquad (H3.3.6)$$

The "true" value is expected to be within $019 \leq \gamma \leq 0.195$. For further computations $\gamma \approx 0.19$ was used. The parameter controlling the effect of swelling stresses on diffusivity is in the following considerations expressed as

$$\lambda = \gamma \frac{\varepsilon_0 E}{(1-\nu)} \frac{\Delta V_w}{R\Theta}. \qquad (H3.3.7)$$

The diffusion differential equation (H3.3.1) was solved numerically by application of the *Mathematica* procedure *NDSolve* [H3.16]. Using the properties at room temperature: $E = 73$ GPa, $\nu = 0.17$, $\varepsilon_0 \cong 0.13$, $\Delta V_w \cong 15$ cm^3/mol, yields $\lambda \cong 13$. Diffusion profiles for this value for a room temperature diffusivity of $D_0 = 3 \times 10^{-21}$ m^2/s and $\kappa = 3 \times 10^{-4}$ m$^{1/2}$ are shown in Fig. H3.5a as a function of time t and in Fig. H3.5b for different polar angles φ. In both figures, a plateau can be clearly identified near the tip where the stresses and, consequently, the diffusivity tend to infinity giving rise for an "instantaneous" zone.

Figure H3.5c shows the contour for $C/C_0 = \varepsilon/\varepsilon_0 = \frac{1}{2}$ in a polar diagram by the circles. For $\varphi = 0$, we get the maximum zone extension r_0. The zone height ω, i.e. the maximum in the height coordinate $y = r \sin\varphi$, is given by the condition

$$\omega = \text{Max}[r_{1/2}(\varphi) \times \sin\varphi] \qquad (H3.3.8)$$

The numerical results show a roughly heart-shaped zone contour with the ratio of zone height to length $\omega/r_0 \approx 3/4$.

The zone shape obtained for the case of stress-induced weak phase transformation was derived by McMeeking and Evans [H3.12]

$$r = \frac{8}{\sqrt{27}} \omega \cos^2(\varphi/2) \qquad (H3.3.9)$$

This dependency is introduced in Fig. H3.5c as the continuous curve computed for the same value of r_0 as was numerically obtained. It is obvious that the swelling zone due to water diffusion in the crack-tip stress field is very similar to the zone shape for stress-induced phase transformations in zirconia ceramics.

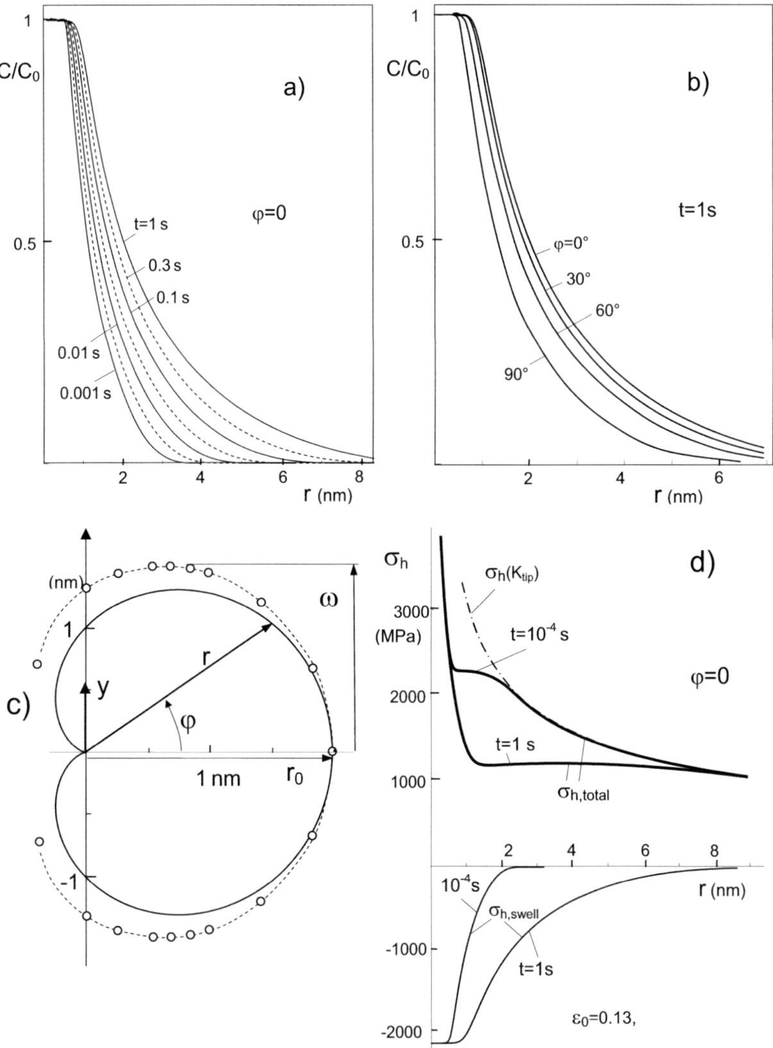

Fig. H3.5 Diffusion and swelling zone: a) profiles as function of time; b) profiles for several angles φ, c) zone shape, circles: results from solution of eq.(H3.3.8), solid curve: zone for phase transformations according to McMeeking and Evans[H3.12] (same r_0 assumed for comparison), d) distribution of hydrostatic stress for $\kappa = 3\times10^{-4}\,m^{1/2}$ (K_{tip}=0.32 MPa√m) after $t=10^{-4}$ and 1s.

Finally, Fig. H3.5d illustrates the distribution of the total hydrostatic stress for $t = 1$ s and $t = 10^{-4}$s as the bold curves and the hydrostatic swelling stress $\sigma_{h,swell}$ by the thin curves.

The dash-dotted curve $\sigma_h(K_{tip})$ represents the hydrostatic stress caused exclusively by the stress intensity factor K_{tip} according to eq.(H3.2.6). Superimposition of $\sigma_h(K_{tip})$ and the negative hydrostatic swelling stress $\sigma_{h,swell}$ (thin solid curves) $\sigma_{h,total} = \sigma_h(K_{tip}) + \sigma_{h,swell}$ results in the thick solid curves exhibiting roughly plateaus of the total stress $\sigma_{h,total} \approx$ constant (thick solid curve).

H3.4 Swelling zone and shielding stress intensity factor

For the computation of shielding stress intensity factors, an effective zone size has to be computed. This zone height, ω_{eff}, is obtained as [H3.15]

$$\omega_{eff} = \left(\frac{1}{2C_0} \int_0^\infty \frac{C(\omega')}{\sqrt{\omega'}} d\omega' \right)^2 \tag{H3.4.1}$$

This zone height differs slightly from the ω defined by eq.(H3.3.8).

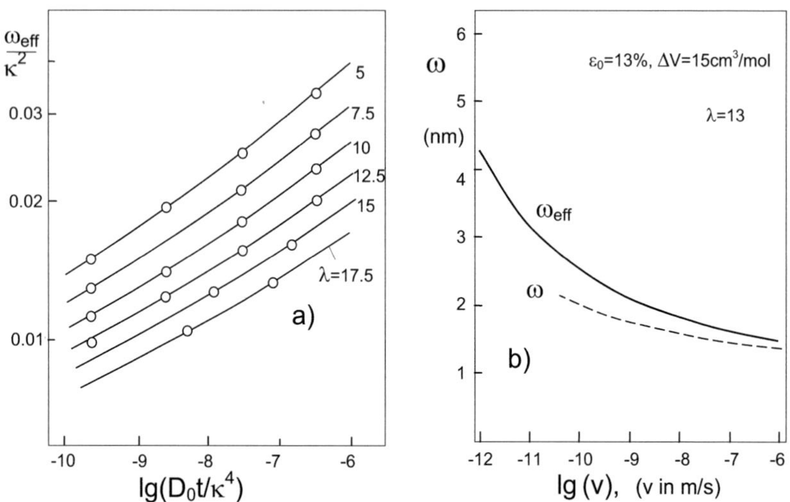

Fig. H3.6 a) Effective zone thickness ω_{eff} computed by eq.(H3.4.1) for a non-growing crack, parameter λ defined by (H3.3.7), **b)** effective zone height for a growing crack as a function of crack rate v.

Results are shown in Fig. H3.6a by the circles. The numerical values including swelling stresses show the same trend as visible from eq.(H3.2.7). Therefore, the numerical results for $\kappa/(D_0 t)^{1/4}<10^{-6}$ were expressed by a similar relation

$$\omega_{eff} \cong \frac{q\,\kappa^2}{\left(\text{PLog}\left[\dfrac{\kappa\,p}{(D_0\,t)^{1/4}}\right]\right)^2} \tag{H3.4.2}$$

with free parameters q and p. These values were varied systematically to achieve the best agreement with the numerical data. For this purpose, the Mathematica routine "NonlinearFit" [H3.16] was applied.

From this procedure it results a constant value for q. The parameter p as a function of λ could be approximated for $0<\lambda\leq17.5$ by an exponential function. The result is

$$q = 0.225\,,\quad p \cong 0.45\exp(0.13\lambda) \tag{H3.4.3}$$

The fit results are introduced in Fig. H3.6a as the curves. For $\lambda=13$ the ratio of ω according to (H3.3.8) and ω_{eff} was found to be smaller than 1, namely, for $t=1$s: $\omega/\omega_{eff}=0.84$ and for $t=10^{-3}$s: $\omega/\omega_{eff}=0.91$ (Fig. H3.6b).

In the case of a crack growing at the constant rate v, the time available for the formation of the swelling zone is roughly $t \approx \omega_{eff}/v$. This results in

$$\omega_{eff} = \frac{0.9\,\kappa^2}{\left(\text{PLog}\left[\dfrac{4p^2}{\sqrt{0.9}}\kappa\sqrt{\dfrac{v}{D_0}}\right]\right)^2} \tag{H3.4.4}$$

By introducing ω_{eff} into eq.(H3.2.1) instead of b the related shielding stress intensity factor reads

$$K_{sh} = -0.209\frac{\varepsilon_0 E}{1-v}\frac{\kappa}{\text{PLog}\left[\dfrac{4p^2}{\sqrt{0.9}}\kappa\sqrt{\dfrac{v}{D_0}}\right]} \tag{H3.4.5}$$

This shielding stress intensity factor gives rise for the same presentation in terms of R-curves as shown in Section H2.2.2 for the case of ion exchange.

The effect of the volume strain ε_0 and the activation volume ΔV on the effective zone height is shown in Fig. H3.7. The influence on K_{sh} is represented in Fig. H3.8. The bold curves always indicate the parameter set used in [H3.15]. It should be noted that for the dashed curves the fitting results of Fig. H3.6a had to be strongly extrapolated.

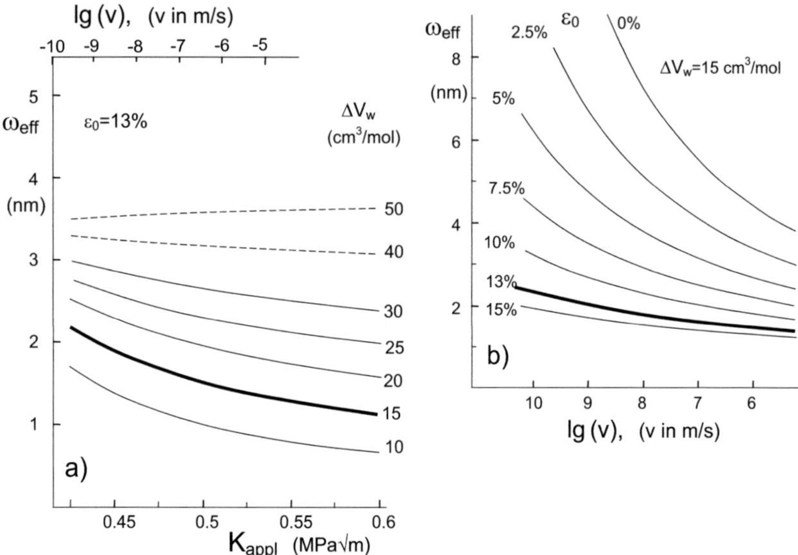

Fig. H3.7 Effective zone thickness ω_{eff} computed by eq.(H3.4.4) for a growing crack, dashed curves strongly extrapolated, a) effect of the activation volume, b) effect of the swelling strain $\varepsilon_0=\varepsilon_v(r{=}0)$.

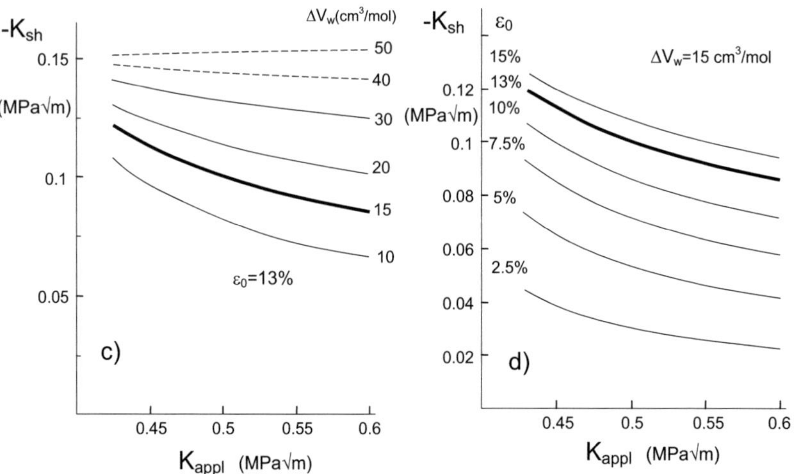

Fig. H3.8 Shielding stress intensity factor K_{sh} computed by eq.(H3.4.5) for a growing crack, a) effect of the activation volume, b) effect of the swelling strain ε_0.

H3.5 Estimation of swelling strain from surface overlapping

Atomic force microscope (AFM) measurements on fracture surfaces from [H3.17] were used to estimate the swelling strains [H3.3]. From Fig. H3.9a one can see that there is extra material (the dark grey area) at the crack tip that was not there before the crack was formed in the glass and held in water at $K_I = 0.254$ MPa·m$^{1/2}$ for 80 days. If the crack were just loaded and unloaded, the crack profile would be relatively flat and to experimental precision, the two fracture surfaces would overlap perfectly.

Based on these considerations, the excess material ahead of the actual crack tip can be interpreted as a water diffusion zone where swelling by the volumetric strain ε_0 took place. After complete cracking of the specimen, each half of this region can expand normally to the new crack face. This results in a volume expansion and in an overlapping of the surface profiles. Figure H3.9b gives measurements of the overlapping displacements with a maximum value of about 6 nm.

In order to study the swelling effect, a finite element (FE) modelling of the heart-shaped zone was performed for a constant volume strain. The displacements are shown in normalized representation in Fig. H3.9c. The maximum displacement from FE-modelling was found at 0.6ω ahead of the crack tip to be $v_y/(\omega\varepsilon_v)\cong0.55$ for $\varepsilon_v=\varepsilon_0$=const. From Fig. H3.9b and H3.9c one can conclude that $\omega\varepsilon_0 \cong 11\,\text{nm}$.

The width at half of maximum displacement, $x_{1/2}/\omega$, is $x_{1/2}/\omega$=1.8 (see Fig. H3.9c). The corresponding measured value of $x_{1/2}$ is $x_{1/2} \approx 140$ nm (see Fig. H3.9b). From these two results we obtain

$$\omega \cong 80\ \text{nm} \quad \Rightarrow \quad \varepsilon_v = 13.7\% \qquad\qquad (\text{H3.5.1})$$

(i.e. a linear strain of about 4.6%).

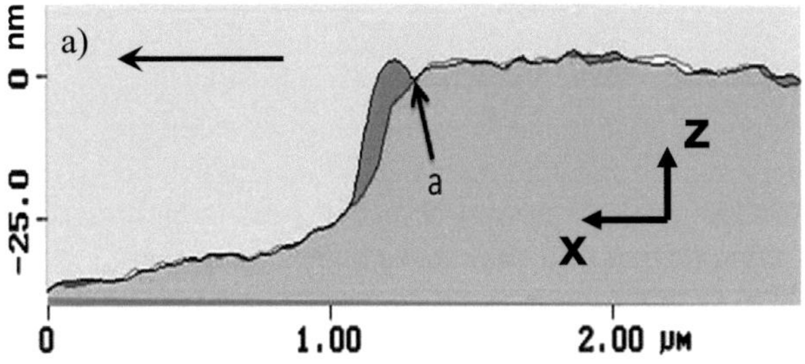

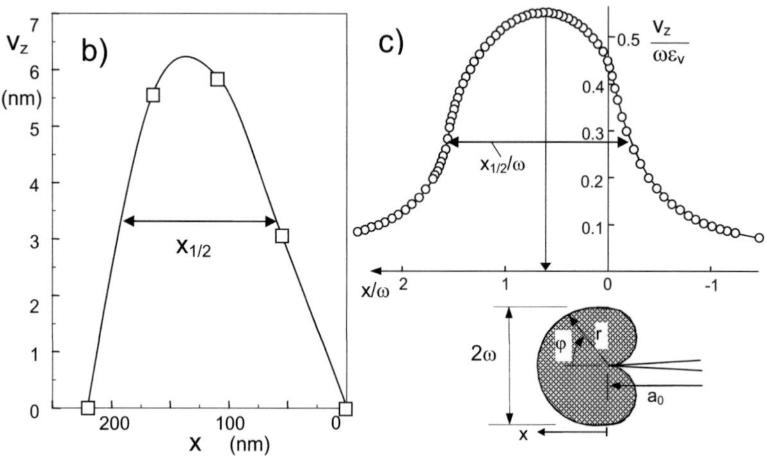

Fig. H3.9 a) Overlapping of crack faces for an arrested crack after final fracture [H3.15, H3.17]. The arrow gives the direction of crack growth, b) crack-face displacement, V_z, from a), c) finite element result for the surface displacements due to a volumetric swelling strain ε_v, resulting in a heart-shaped zone.

H3.6 Experimental attempts for the determination of K_{sh}

The preceding sections showed the basic effects, which ensure the existence of a shielding stress intensity factor K_{sh} due to a stress-enhanced water diffusion zone near a crack tip in silica accompanied by volume swelling. The theoretical results in [H3.3, H3.15] were obtained with an activation volume for stress-enhanced diffusion of $\Delta V_w=$ 15 cm^3/mol and a maximum swelling strain at the surface ε_0=13% by solving the 2-dimensional diffusion equation. The desired result in the end of the shielding studies is the relation between the shielding stress intensity factor and the externally applied one, K_{sh}=f(K_{appl}), which allows the measured material properties to be expressed in terms of the true crack-tip stress intensity factor, e.g. $v(K_{appl})$ by $v(K_{tip})$.

There are several unpredicticabilities as the influence of the stress-enhanced volume swelling strain ε_0 as a function of humidity and temperature and of the exact value ΔV_w. In addition the accuracy of measured diffusivities has to be questioned to a certain extend since even in unstressed silica surfaces compressive swelling stresses will be generated which again affect the diffusion via eq.(H3.1.3). Therefore, the measured diffusivities are somewhat smaller than D_0. Such an influence is briefly discussed in the Appendix I4.

In order to compare the theoretical predictions with real material behavior, the K_{sh}-K_{appl}-dependency can be measured directly as a function of temperature and humidity of the environment. This comparison possibly may result in imroved parameters ε and ΔV_w. Four possible tests are proposed.

H3.6.1 Measurement of crack opening displacements

One of the actually studied methods may be outlined here in some detail [H3.18]. For this purpose, a crack in a DCB- or a DCDC-specimen is kept at a certain constant applied stress intensity factor resulting in the crack-growth rate v. After a sufficiently large crack extension of $\Delta a > 5$ ω_{eff}, the applied stress intensity factor is continuously reduced and the decreasing crack opening displacement δ observed as was done by Michalske and Fuller [H3.19] in their early crack-healing study via optical interferences. Since the crack opening displacements due to swelling are predominantly present in crack-tip distances of a few zone heights, these measurements require methods with very high resolution. Possibly, also for this purpose AFM-methods might be applicable as were used recently by Pallares et al. [H3.20] for the determination of displacements due to the Laplace pressure in the condensate zone at the crack tip.

For the total near-tip displacements δ_{tip} in the crack-tip distance r it holds

$$\delta_{tip} = \delta_{appl} + \delta_{sh} = \sqrt{\frac{8}{\pi}} \frac{K_{tip}}{E'} \sqrt{r} = \sqrt{\frac{8}{\pi}} \frac{K_{appl} + K_{sh}}{E'} \sqrt{r} \qquad \text{(H3.6.1)}$$

At the load where the measurable displacements disappear, $\delta_{\text{tip}}=0$, it results from eq.(H3.6.1):

$$\delta_{tip} = 0 \Rightarrow K_{tip} = K_{appl} + K_{sh} = 0 \Rightarrow K_{sh} = -K_{appl} \qquad \text{(H3.6.2)}$$

The applied stress intensity factor at disappearing displacements can be computed and, consequently, K_{sh} has been measured. The unloading has not necessarily to be done in the testing device (DCDC, DCB) and at the test temperature since the stress intensity factor caused by swelling is independent on the applied load and remains unchanged even after unloading. The closure measurement may for instance be carried out in a vacuum chamber in order to avoid a superimposed Laplace-pressure.

The crack-closure situation is represented in Fig. H3.10 in terms of the displacements. In this example it is chosen $\omega=2$ nm and $K_{\text{sh}}= -0.1$ MPa$\sqrt{\text{m}}$. In Fig. H3.10a the displacements under the condition $K_{\text{tip}}=0$ are plotted for the region very close to the tip.

The displacements for a swelling zone of height ω with the shielding stress intensity factor K_{sh} is given by [H3.21]

$$\delta_{swell} = \delta_{sh} \cong \frac{K_{sh}\sqrt{\omega}}{E'} A \tanh^{1/p}\left(\frac{1}{A}\sqrt{\frac{8}{\pi}}\sqrt{\frac{r}{\omega}} \right)^p, \quad A = 2.2, \ p = \frac{6}{7} \qquad \text{(H3.6.3)}$$

Figure H3.10b shows the same curves over an extended region. It becomes clear from Fig. H3.10a that the shielding displacements δ_{sh} reach a constant value of about 0.13 nm already after about $r=10$ nm. Since according to Fig. H3.10 the crack opening displacements due to swelling are predominantly present in crack-tip distances of a few zone heights, their measurement needs methods with very high resolution. Possibly, also for this purpose AFM-methods might be applicable.

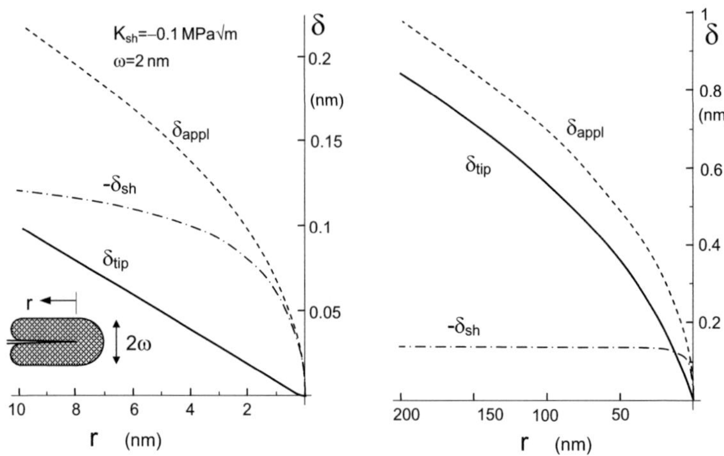

Fig. H3.10 Displacements at the crack-closure condition $K_{tip}=0$.

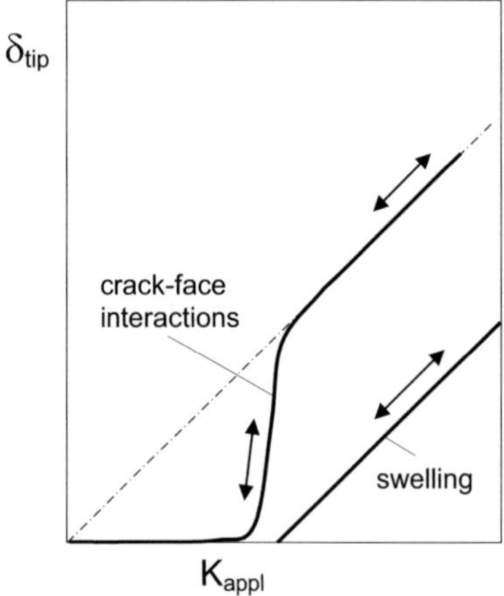

Fig. H3.11 COD-behaviour during unloading and re-loading of a crack caused by crack-face interactions and crack-tip swelling.

The different effects for the crack-closure can be separated experimentally, either by removing the surface water contamination or by recording the displacement as a function of load. Whereas the crack-face interactions via water layers can develop only at the smallest displacements, the swelling effect is independent on the displacement. This different behaviour is schematically illustrated in Fig. H3.11.

H3.6.2 Measurement of fracture toughness after subcritical crack extension

The occurrence of a swelling zone at the crack tip after subcritical crack growth will affect the toughness and strength. Measurements should result in an apparent fracture toughness K^*

$$K^* = K_{Ic} - K_{sh} \geq K_{Ic} \qquad (H3.6.4)$$

A determination of K_{sh} is then possible by the following steps:

- A crack is propagated in water at a certain crack-growth rate in a fracture mechanics test specimen, e.g. a Double Cantilever Beam test. After a crack extension Δa of at least 3-5 times the expected zone height ω, the specimen is unloaded.

- Then, the specimen fried for instance in vacuum.

- A second fracture toughness test is carried out in the same way. Then, the specimen is annealed in order to remove the swelling stresses. In an inert medium (liquid nitrogen [H3.22]) the apparent toughness K^* is determined. Due to eq. (H3.6.4) K_{sh} is known.

Since K_{appl} during the slow crack growth test is known from the observable crack length and the applied load, the related K_{tip} is given as $K_{tip} = K_{appl} + K_{sh}$.

H3.6.3 Measurement of inert strength after subcritical crack extension

Swelling zones at crack tips must also affect the inert strength of silica. This fact allows the shielding stress intensity factor to be determined. For the evaluation of K_{sh}, a large number of N specimens are tested in 3 series of inert strength tests. Before the experiments, all samples may be annealed at 1140°C in order to remove residual stresses caused by the surface treatment. The first series of $N/3$ specimens is fractured under inert conditions in liquid N_2. The results for the initial flaw distribution, $\sigma_{c,0}$, may be schematically represented in the Weibull plot of Fig. H3.12 as the dashed curve. The remaining $2/3N$ specimens are loaded with a static load σ_1 in a humid environment for a time t_1. The size of the cracks extends by subcritical crack growth. Depending on the height of the applied load, a part of the specimens may fail during the time t_1. The survivals with their increased crack size must then show swelling zones in the crack-propagation region. A part of $N/3$ specimens is annealed again in order to remove the swelling stresses, which are responsible for crack-tip shielding. The results of inert strength tests on these samples, $\sigma_{c,1}$, are indicated by the dash-dotted curve in Fig. H3.12. The remaining $N/3$ specimens are also fractured under inert conditions. This portion of samples with strengths $\sigma_{c,2}$ is shown as the solid curve. The difference between the solid and the dash-dotted curves reflects the effect of the swelling zone. The amount of crack extension Δa can simply be obtained from the difference of the dash-dotted and the dashed curves as

$$\Delta a = \left(\frac{K_{Ic}}{\sigma_{c,1}Y}\right)^2\left(1 - \frac{\sigma_{c,1}^2}{\sigma_{c,0}^2}\right) \tag{H3.6.5}$$

with $Y \cong 1.3$ for a semi-circular crack shape. If the crack increment Δa is clearly larger than the height ω of the swelling zone (expected to be less 1 µm), the full shielding term is reached. Finally, the shielding stress intensity factor results from

$$K_{sh} = K_{Ic}\left(1 - \frac{\sigma_{c,2}}{\sigma_{c,1}}\right) \tag{H3.6.6}$$

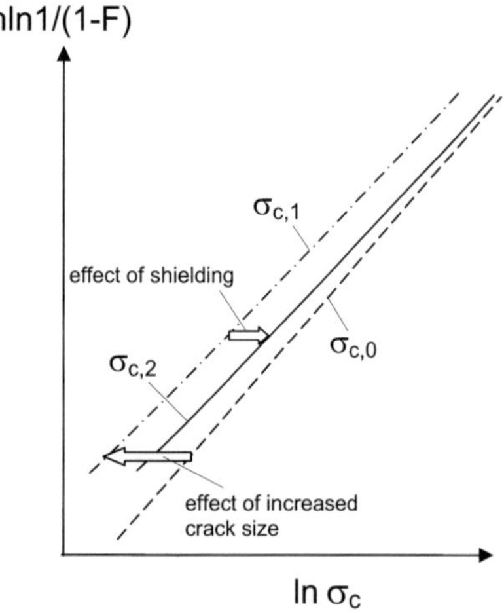

Fig. H3.12 Schematic Weibull representation of inert strengths σ_c expected for silica; dashed line: inert strengths corresponding to the initial surface flaw population, dash-dotted curve: inert strengths after previous subcritical crack propagation with the expected swelling stresses removed by annealing, solid curve: inert strengths including the swelling stresses.

H3.6.4 Measurement of inert strength of artificially damaged specimens after subcritical crack extension

The previously mentioned strength tests are quite expensive with respect to the number of samples because a statistical evaluation is necessary. A greatly reduced effort is possible applying artificial surface cracks. For Knoop indentation cracks a possible procedue would consist of the following steps:

- A few initially annealed specimens are damaged by the indentation procedure.
- The residual stress zone developed below the contact area (about 30-40% of the crack depth) is removed by grinding. The additionally introduced grinding cracks are without importance for fracture so far the Knoop indentation crack is the largest crack in the specimen.
- The specimens are then annealed in order to remove remaining residual stresses.
- After a constant load test, in which the swelling zone is generated, the specimens are fractured in liquid N_2.
- From the inert strength σ_c and the dimensions of the crack, the apparent fracture toughness K^* is computed via

$$K^* = \sigma_c Y \sqrt{a} \qquad\qquad (H3.6.7)$$

and K_{sh} follows from eq.(H3.6.4).

The shape of the Knoop indentation cracks after removal of the residual stress zone may differ from the semi-circular shape. The geometric function Y has then to be taken from a number of original papers or fracture mechanics handbooks. For semi-elliptic surface cracks see e.g. [H3.23, H3.24] and sections of circular cracks [H3.24]. If the crack shape cannot be described with sufficient accuracy by existing stress intensity factor solutions, a few additional tests on annealed specimens with removed swelling stresses are recommended followed by an evaluation via eq.(H3.6.6).

Final Remark:

Very similar to the effect of ion-exchange layers (Section H2), the increase of the shielding term by swelling is an effect on the length scale of a few nm. Saturation is reached after crack extensions of less than $1\mu m$. From this point of view, also silica can be considered as a material without an R-curve effect in the usual sense, i.e. for crack extensions of Δa in the order of μm to mm.

References H3

H3.1 Zouine, A., Dersch, O., Walter, G., Rauch, F., Diffusivity and solubility of water in silica glass in the temperature range 23-200°C, Phys. Chem. Glasses, **48** (2007), 85-91.

H3.2 Nogami, M., Tomozawa, M., Effect of stress on water diffusion in silica glass, J. Am. Ceram. Soc., **67**(1984), 151-154.

H3.3 S. M. Wiederhorn, T. Fett, G. Rizzi, S. Fünfschilling, M.J. Hoffmann, J.-P. Guin, Effect of Water Penetration on the Strength and Toughness of Silica Glass, J. Am. Ceram. Soc. **94** [S1] (2011) S196-S203.

H3.4 Gorbacheva, M.I., Zaoints, R.M., Moisture expansion of various crystalline and amorphous phases, Steklo I Keramika, **11**(1974), 18-19.

H3.5 Thurn, J., Water diffusion coefficient measurements in deposited silica coatings by the substrate curvature method, J. Non-Cryst Solids, **354**(2008), 5459-65.

H3.6 Kuschke, W.M., Carstanjen, H.D., Pazarkas, N., Plachke, D.W., Arzt, E., Influence of water absorption by silicate glass on the strains in passivated Al conductor lines, J. of Electronic Mater. **27**(1998), 853-857.

H3.7 Oka, Y., Wahl, J.M., Tomozawa, M., Effect of surface energy on the mechanical strength of a high-silica glass, J. Am. Ceram. Soc. **64**(1981), 456-460.

H3.8 Shelby, J.E., Density of vitreous silica, J. Non-Cryst. **349**(2004), 331-336.

H3.9 Bruckner, R., Glastech. Ber. **43**(1970), 8.

H3.10 Bruckner, R., J. Non-Cryst. **5**(1971), 281.

H3.11 Shackelford, J.F., Masaryk, J.S., Fulrath, R.M., J. Am. Ceram. Soc. **53**(1970), 417.

H3.12 McMeeking, R.M., Evans, A.G., Mechanics of Transformation-Toughening in Brittle Materials, J. Am. Ceram. Soc. **65**(1982), 242-246.

H3.13 Evans, A.G., Faber, K.T., Crack-growth resistance of microcracking brittle materials, J. Am. Ceram. Soc. **67**(1984), 255-260.

H3.14 Budiansky, B., Hutchinson, J.W., Lambropoulos, J.C., Continuum theory of dilatant transformation toughening in ceramics, Int. J. Solids and Structures, **25**(1989), 635-646.

H3.15 S. M. Wiederhorn, T. Fett, G. Rizzi, M.J. Hoffmann, J.-P. Guin, The Effect of Water Penetration on Crack Growth in Silica Glass, submitted to Engng. Fract. Mech.

H3.16 *Mathematica*, Wolfram Research, Champaign, USA.

H3.17 J.-P. Guin and S.M. Wiederhorn, Investigation of the Subcritical Crack Growth Process in Glass by Atomic Force Microscopy, Ceram. Trans. **199** 13-21 (2007).

H3.18 S. M. Wiederhorn, T. Fett, S. Wagner, M.J. Hoffmann, J.-P. Guin, unpublished.

H3.19 T.A. Michalske, E.R. Fuller, Closure and repropagation of healed cracks in silicate glass, *J. Am. Ceram. Soc.* **79**[1] 51-57 (1996)

H3.20 Pallares, G., Grimaldi, A., George, M., Ponson, L., Ricotti, M., Quantitative analysis of crack closure driver by Laplace pressare in silica glass, J. Am. Ceram. Soc. **94**(2011), 2613-18.

H3.21 Fett, T., Guin, J.P., Wiederhorn, S.M., Interpretation of effects at the static fatigue limit of soda-lime-silicate glass, Engng. Fract. Mech. **72**(2005), 2774-279.

H3.22 S.M. Wiederhorn, Fracture surface energy of glass, J. Am. Ceram. Soc. **52**(1969), 99-105.

H3.23 Newman, J.C., Raju, I.S., An empirical stress intensity factor equation for the surface crack, Engng. Fract. Mech. **15**(1981), 185-192.

H3.24 S. Strobl, R. Danzer, P. Supancic, T. Lube, Surface crack in tension or in bending - A reassessment of the Newman and Raju formula in respect to fracture toughness measurements in brittle materials, to appear in J. Eur. Ceram. Soc.

APPENDIX

I1 Stress intensity factors for indentation cracks

I1.1 Application of averaged stress intensity factors

A simple fracture mechanics tool for the determination of the shape of semi-elliptical surface cracks is the computation of so-called averaged stress intensity factors. By application of this weight function procedure it is possible to determine stress intensity factor on the basis of a relation by Rice [I1.1] which relates the variation of the reference crack opening displacement v_r in a certain reference loading case, e.g. σ_r=const., to the stress intensity factor in the actual loading case σ

$$\frac{1}{\Delta S}\int_{(\Delta S)} K_I K_{Ir}\,d(\Delta S) = E'\int_{(S)}\sigma\,\frac{\partial v_r}{\partial(\Delta S)}\,dS \tag{I1.1.1}$$

where $E' = E/(1-\nu^2)$, ν is Poisson's ratio, K_{Ir} is the reference stress intensity factor, and ΔS a virtual increment of the crack area. The left-hand side of eq.(I1.1.1) gives rise for defining the stress intensity factor $\overline{K}_I$ according to

$$\overline{K}_{Ir} = \sqrt{\frac{1}{\Delta S}\int_{(\Delta S)} K_{Ir}^2\,d(\Delta S)} \tag{I1.1.2}$$

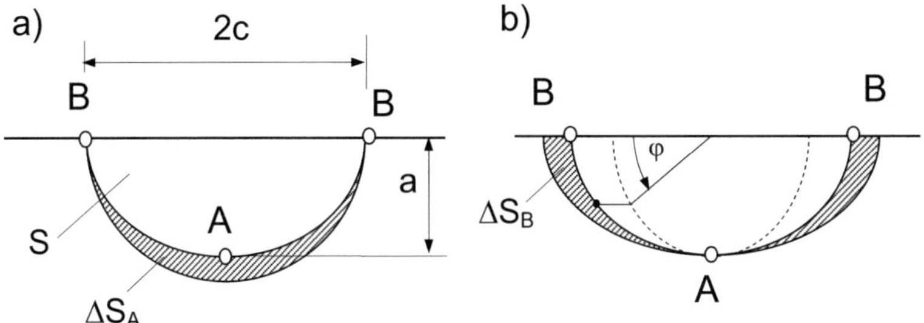

Fig. I1.1 Virtual crack extensions according to Cruse and Besuner [I1.2].

The stress intensity factor according to the right-hand side of eq.(I1.1.1) can be computed from

$$\overline{K} = \frac{E'}{\overline{K}_r}\int_{(S)}\sigma\,\frac{\partial v_r}{\partial(\Delta S)}\,dS \tag{I1.1.3}$$

The derivative $\partial v_r / \partial(\Delta S)$ is called the weight function which of course depends on the specially chosen crack area increment ΔS. Fett and Munz [11.3] gave a procedure for the determination for the reference crack opening displacement field of semi-elliptic surface cracks. Cruse and Besuner [11.2] could show that this type of stress intensity factor is related to the energy release rate released during a virtual crack surface extension of $\partial(\Delta S)$. This holds at least for not too abnormal stress distributions (see e.g.[11.3]).

Cruse and Besuner [11.2] proposed two independent virtual crack changes that preserve a semi-elliptical crack shape, namely, crack depth increment Δa with width c kept constant (Fig. 11.1a)

$$\Delta S_A = \tfrac{1}{2}\pi c \Delta a \qquad (11.1.4)$$

or crack width increment Δc with depth $a = \text{const}$

$$\Delta S_B = \tfrac{1}{2}\pi a \Delta c \qquad (11.1.5)$$

as illustrated in Fig. 11.1b by the hatched areas.

The stress intensity factors are represented by the geometric functions F_A (for a virtual crack increment at point A) and F_B (for an increment Δc at the surface)

$$\overline{K}_{A,B} = \sigma_0 \sqrt{\pi a}\, F_{A,B}\,, \qquad (11.1.6)$$

where σ_0 is a characteristic stress (e.g. a remote tensile stress for tensile loading or the outer fibre bending stress in bending).

It should be emphasized that the Cruse-Besuner approach for the computation of averaged stress intensity factors was examined intensively in the eighties and early nineties. Only a few papers may be mentioned in this context, which show that the method is applicable to various cracks. The theoretical and experimental analyses on simple semi-elliptical surface cracks in plates and bars (see e.g. [11.4, 11.5, 11.6]) were extended to more complicated crack problems as for instance almond- and sickle-shaped cracks in rods [11.5] and corner cracks [11.7].

The use of this type of stress intensity factors allowed very good predictions for the development of the crack shape in tension and bending and correct predictions of crack growth rates in fatigue tests. In the following considerations, only stress intensity factors defined by (11.1.1) and (11.1.3) will be addressed. For reasons of simplicity the bar above K is dropped.

I1.2 Applied stress intensity factors for indentation cracks

I1.2.1 Stress intensity factors for remote tension

Geometric functions F_A and F_B according to (I1.1.6) are for $0.7 \leq a/c \leq 1$ and $0 \leq a/t \leq 0.1$

$$F_A = 1.118 - 0.6127\frac{a}{c} + 0.1609\left(\frac{a}{c}\right)^2 + 0.8449\left(\frac{a}{t}\right)^2 - 0.7257\left(\frac{a}{c}\right)\left(\frac{a}{t}\right)^2 \qquad (I1.2.1)$$

$$F_B = 0.7683 - 0.00746\frac{a}{c} - 0.0534\left(\frac{a}{c}\right)^2 + 0.7967\left(\frac{a}{t}\right)^2 - 0.5529\left(\frac{a}{c}\right)\left(\frac{a}{t}\right)^2 \qquad (I1.2.2)$$

$$\frac{F_B}{F_A} \cong 0.6421 + 0.531\frac{a}{c} - 0.111\left(\frac{a}{c}\right)^2 + 0.0022\left(\frac{a}{t}\right)^2 + 0.16\left(\frac{a}{c}\right)\left(\frac{a}{t}\right)^2 \qquad (I1.2.3)$$

with the thickness t (in contrast to one-dimensional cracks for which the thickness is mostly abbreviated by W). The influence of a/t representing the influence of the free rear wall is very small for pure tension.

The aspect ratio of the equilibrium ellipse (at which $K_A = K_B$) is

$$\left(\frac{a}{c}\right)_{eq} \cong 0.812 - 0.350\left(\frac{a}{t}\right)^2 \qquad (I1.2.4)$$

For $(a/c)_{eq}$ a maximum influence of less than 1% for $a/t \leq 0.15$ can be stated with the related stress intensity factor variation less than 0.35%, negligible in practice.

The data for an extended region of $0.25 \leq a/c \leq 1$ can be approximated for $a/t = 0$ by

$$F_A \cong 1.191 - 0.7876\frac{a}{c} + 0.2582\left(\frac{a}{c}\right)^2 \qquad (I1.2.5)$$

$$F_B \cong 0.699 + 0.1541\frac{a}{c} - 0.149\left(\frac{a}{c}\right)^2 \qquad (I1.2.6)$$

I1.2.2 Stress intensity factors for bending load

Stress intensity factor solutions for cracks in bars under bending load are available in fracture mechanics literature (see e.g. [I1.8]). The geometric functions F_A and F_B according to (I1.1.6) were determined in [I1.3, I1.10] by use of these results as the reference stress intensity factors. For a wide aspect ratio range of $0.5 \leq a/c \leq 1$ and relative crack depths of $0 \leq a/t \leq 0.15$, they read

$$F_A = 1.107 - 0.5845\frac{a}{c} + 0.1436\left(\frac{a}{c}\right)^2 - 1.135\frac{a}{t} + 0.8454\left(\frac{a}{t}\right)^2$$
$$+ 0.311\frac{a}{c}\frac{a}{t} - 0.7621\left(\frac{a}{c}\right)\left(\frac{a}{t}\right)^2 \tag{I1.2.7}$$

$$F_B = 0.755 + 0.0232\frac{a}{c} - 0.0714\left(\frac{a}{c}\right)^2 - 0.569\frac{a}{t} + 0.739\left(\frac{a}{t}\right)^2$$
$$+ 0.0528\frac{a}{c}\frac{a}{t} - 0.541\left(\frac{a}{c}\right)\left(\frac{a}{t}\right)^2 \tag{I1.2.8}$$

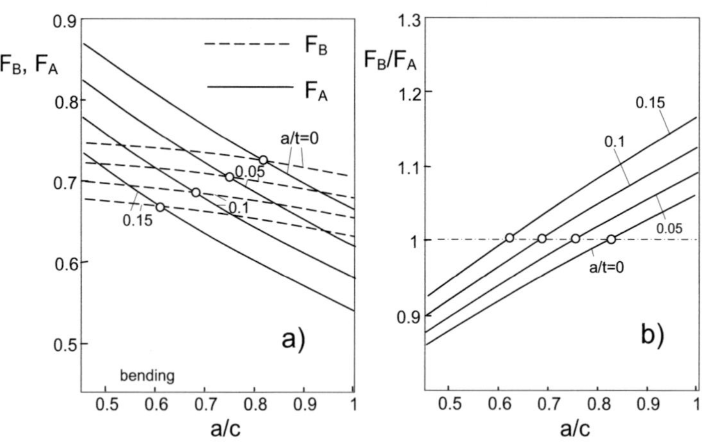

Fig. I1.2 Semi-elliptical surface crack under bending load; a) influence of the relative crack depth a/t and the aspect ratio a/c on the stress intensity factors, b) ratio of stress intensity factors, (circles indicate $F_A=F_B$).

The geometric functions and the ratio of F_A/F_B are plotted in Fig. I1.2. Identical stress intensity factors at points A and B are indicated by the circles. For the aspect ratio $(a/c)_{eq}$ of the equilibrium ellipse it approximately holds at small depths $a/t<0.25$

$$\left(\frac{a}{c}\right)_{eq} \cong 0.812 - 1.382\frac{a}{t} + 0.1344\left(\frac{a}{t}\right)^2 \tag{I1.2.9}$$

or in terms of the simpler measurable term c/t

$$\left(\frac{a}{c}\right)_{eq} = \frac{1}{(c/t)^2}\left[3.72 + 5.141\frac{c}{t} - 4.516\sqrt{0.6787 + 1.876\frac{c}{t} + \left(\frac{c}{t}\right)^2}\right] \tag{I1.2.10}$$

For practical applications with a manageable number of terms, the geometric functions in the ranges of $0.5 \leq a/c \leq 1$ and $0 \leq a/t \leq 0.15$ may be simplified as

$$F_A \approx 1.02 - 0.36\frac{a}{c} - 0.866\frac{a}{t} \qquad (I1.2.11)$$

$$F_B \approx 0.79 - 0.081\frac{a}{c} - 0.479\frac{a}{t} \qquad (I1.2.12)$$

I1.3 Residual stress intensity: Knoop indentation cracks

During the loading/unloading procedure by a Knoop indenter, a residual stress zone develops below the contact area. According to the model proposed by Marshall [I1.9] a prolate spheroid was chosen for the shape of the irreversibly deformed 'plastic' zone with the major axis b_1 and the depth b_2 (Fig. I1.3).

If σ_{res} is the residual stress assumed to be constant over the semi-elliptic cross section with the half-axes b_1 and b_2, the total force normal to the crack plane is

$$P_{res} = \tfrac{1}{2}\sigma_{res}\pi\, b_1 b_2 \qquad (I1.3.1)$$

The stress intensity factors may be scaled with this force as

$$K_{A,B} = \frac{2P_{res}}{(\pi a)^{3/2}} F_{A,B} \qquad (I1.3.2)$$

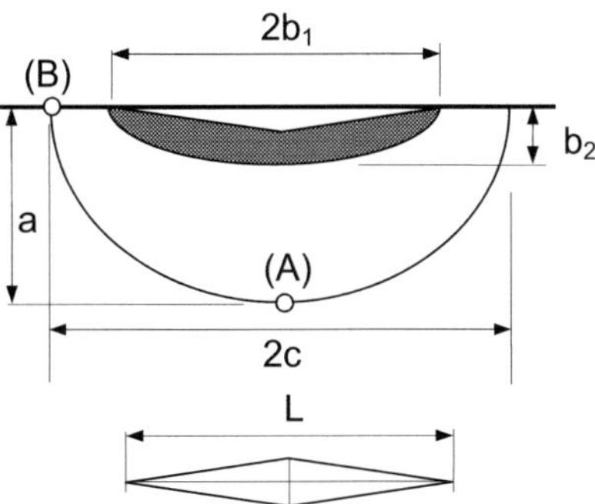

Fig. I1.3 Semi-elliptical crack loaded by a residual stress zone of length $2b_1$ and width b_2 as suggested by Marshall [I1.9].

The geometric functions $F_{A,B}$ of (I1.3.2) are compiled in Tables I1.1 and I1.2 (for additional results see [I1.10]). For any parameter in the ranges of $1 \leq c/b_1 \leq \infty$, $1 \leq b_1/b_2 \leq \infty$, and $0.7 \leq a/c \leq 1.2$, F_B can be approximated by

$$F_B \cong -0.46 + 1.8\alpha + \left(0.55 + 0.531\alpha\right)\exp[-(2.11 + 0.186\alpha)\sqrt{\frac{c}{b_1} - 0.97}] -$$
$$-\frac{1}{4}\left(\frac{b_1}{c}\right)^2\left(\frac{1}{\beta} - \frac{1}{3}\right) \tag{I1.3.3}$$

with the abbreviations $\alpha = a/c$ and $\beta = b_1/b_2$.

For $1.1 \leq c/b_1 \leq \infty$, $1.5 \leq b_1/b_2 \leq \infty$, and $0.7 \leq a/c \leq 1$, F_A reads

$$F_A = 0.51 + 1.17\alpha - 0.645\alpha^2 - (1.317 + 6.18\alpha - 26.9/\beta^2)\exp[-3.7\sqrt{c/b_1}] \tag{I1.3.4}$$

c/b_1	$a/c=0.7$	0.8	0.9	1	1.1	1.2	1.3
0.8	1.506	1.728	1.941	2.146	2.342	2.531	2.711
0.9	1.565	1.789	2.007	2.217	2.419	2.614	2.801
1.0	1.407	1.621	1.830	2.033	2.229	2.418	2.601
1.1	1.207	1.410	1.609	1.802	1.989	2.170	2.344
1.2	1.118	1.315	1.509	1.698	1.882	2.059	2.231
1.3	1.059	1.254	1.445	1.631	1.812	1.988	2.158
1.4	1.017	1.209	1.398	1.583	1.763	1.937	2.105
1.6	0.985	1.176	1.363	1.546	1.725	1.898	2.065
1.8	0.924	1.111	1.296	1.476	1.653	1.824	1.990
2.0	0.899	1.084	1.268	1.448	1.623	1.794	1.959
2.2	0.881	1.065	1.248	1.427			
2.4	0.867	1.051	1.233	1.412			
2.6	0.857	1.040	1.221	1.399			
3.0	0.842	1.024	1.205	1.383			
3.5	0.830	1.011	1.192	1.369			
4.0	0.822	1.003	1.183	1.361			
4.5	0.817	0.998	1.178	1.355			
∞	0.802	0.980	1.158	1.336			

Table I1.1 Normalized stress intensity factor F_B for $b_1 = 3b_2$.

c/b_1	$a/c=0.7$	0.8	0.9	1
0.8	0.874	0.873	0.864	0.849
0.9	0.904	0.909	0.905	0.892
1.0	0.930	0.940	0.939	0.928
1.1	0.947	0.959	0.960	0.950
1.2	0.958	0.973	0.974	0.965
1.3	0.967	0.983	0.985	0.976
1.4	0.973	0.991	0.994	0.985
1.6	0.983	1.002	1.006	0.998
1.8	0.990	1.010	1.014	1.006
2.0	0.994	1.015	1.020	1.012
2.2	0.998	1.019	1.024	1.017
2.4	1.001	1.022	1.027	1.020
2.6	1.003	1.024	1.030	1.022
3.0	1.006	1.028	1.033	1.026
3.5	1.008	1.030	1.036	1.029
4	1.010	1.032	1.038	1.031
4.5	1.011	1.034	1.040	1.032
∞	1.017	1.036	1.041	1.034

Table I1.2 Normalized stress intensity factor F_A for $b_1=3b_2$.

For limited parameter ranges of $0.7 \leq a/c \leq 1$, $1.2 \leq c/b_1 \leq 2$, and $2 \leq b_1/b_2 \leq 6$, the results in [I1.10] can be approximated simply by

$$F_B \approx -0.742 + 1.869\frac{a}{c} + 0.69\frac{b_1}{c} - 0.0904\frac{b_2}{b_1} \tag{I1.3.5}$$

$$F_A \approx 1 + 0.022\left(\frac{a}{c}\right)^2 - 0.116\frac{b_1}{c} + 0.1717\frac{b_2}{b_1} \tag{I1.3.6}$$

I1.4 Residual stress intensity: Vickers indentation cracks

By extending the stress intensity factor solution of [I1.10] to increased ranges of $0 \leq b/a \leq 0.9$ and $0.5 \leq a/c \leq 1$, it results

$$F_{res,A} = 0.4773 + 0.9416\frac{a}{c} + 0.0742\left(\frac{a}{c}\right)^2 - 0.456\left(\frac{a}{c}\right)^3 + 0.4166\frac{a}{c}\left(\frac{b}{a}\right)^2$$
$$+ 0.1145\frac{a}{c}\left(\frac{b}{a}\right)^4 - 0.260\left(\frac{a}{c}\right)^2\left(\frac{b}{a}\right)^2 + 0.2147\left(\frac{a}{c}\right)^2\left(\frac{b}{a}\right)^4 + 0.0944\left(\frac{a}{c}\right)^3\left(\frac{b}{a}\right)^2$$

$$(I1.4.1)$$

$$F_{res,B} = -0.224 + 0.8293\frac{a}{c} + 1.3064\left(\frac{a}{c}\right)^2 - 0.577\left(\frac{a}{c}\right)^3 + 0.5065\frac{a}{c}\left(\frac{b}{a}\right)^2$$
$$- 0.4192\frac{a}{c}\left(\frac{b}{a}\right)^4 - 1.264\left(\frac{a}{c}\right)^2\left(\frac{b}{a}\right)^2 + 0.6835\left(\frac{a}{c}\right)^2\left(\frac{b}{a}\right)^4 + 0.9029\left(\frac{a}{c}\right)^3\left(\frac{b}{a}\right)^2$$

$$(I1.4.2)$$

References I1

I1.1 Rice, J.R., Some remarks on elastic crack-tip stress fields. *Int. J. Solids and Structures* **8**(1972), 751-758.

I1.2 Cruse, T.A., Besuner, P.M., Residual life prediction for surface cracks in complex structural details, J. of Aircraft **12**(1975) 369-375.

I1.3 Fett, T., Munz, D., Stress intensity factors and weight functions, Computational Mechanics Publications, 1997, Southampton.

I1.4 Görner, F., Mattheck, C., Munz, D., Change in geometry of surface cracks during alternating tension and bending, Z. Werkstofftech. **14**(1983), 11-18

I1.5 Caspers, M., Mattheck, C., Munz, D, Propagation of surface cracks in notched and unnotched rods, ASTM STP 1060, pp. 365-389.

I1.6 Mahmoud, M.A., Surface fatigue crack growth under combined tension and bending loading, Engng. Fract. Mech. **36**(1990), 389-395.

I1.7 Varfolomeyev, I.V., Vainshtok, V.A., Krasowsky, A.Y., Prediction of part-through crack growth under cyclic loading, Engng. Fract. Mech. **40**(1991), 1007-1022.

I1.8 Newman, J.C., Raju, I.S., An empirical stress intensity factor equation for the surface crack, Engng. Fract. Mech. **15**(1981), 185-192.

I1.9 Marshall, D.B., Controlled flaws in ceramics: A comparison of Knoop and Vickers indentation, J. Am. Ceram. Soc. **66**(1983), 127-131.

I1.10 Fett T. Stress intensity factors, T-stresses, Weight function (Supplement Volume), IKM55, KIT Scientific Publishing, Karlsruhe; 2009; (open Access: http://digbib.ubka.uni-karlsruhe.de /volltexte/1000013835).

I2 Toughness of single Si$_3$N$_4$ beta-crystals

I2.1 Crack opening displacements

Vickers indentation cracks are a simple tool to determine the angle-dependent fracture toughness of single beta silicon nitride grains. For this purpose, crack paths have to be selected with cracks ending in such a grain. Figure I2.1 illustrates three symmetric crack orientations with respect to the hexagonal crystal axes. If the crack-tip region in which the measurements were carried out is small compared to the crack extension in the grain a' and to the grain ligament d-a', the well-known Irwin parabola enables determining the actual mode-I stress intensity factor K_I from crack opening displacement measured close to a crack tip. The described problem is highly complicated especially in the case of cracks inclined to the crystal axes since the single grain is anisotropic.

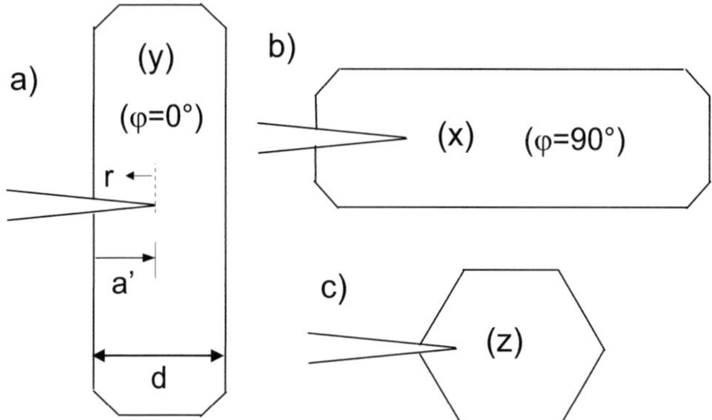

Fig. I2.1 Definition of the crack direction with respect to the c-axis of a single β-crystal.

As can be derived from equations given by Ting [I2.1], the square-root-shaped part of the near-tip crack-surface displacement vector $\mathbf{u}_{tip}$ is related to the stress intensity factor $\mathbf{k}$ by

$$\mathbf{u}_{tip} = \sqrt{\frac{2r}{\pi}}\,\mathbf{L}^{-1}\mathbf{k} \qquad (I2.1.1)$$

where

$$\mathbf{u}_{tip} = (\delta_{I,tip},\delta_{II,tip},\delta_{III,tip})^T, \qquad \mathbf{k} = (K_I,K_{II},K_{III})^T \qquad (I2.1.2)$$

L is the real symmetric second-order tensor in the formalism of Stroh in terms of the elasticity tensor (details in [I2.1]). Since the non-diagonal components of $\mathbf{L}^{-1}$ disappear, i.e. $(\mathbf{L}^{-1})_{13}=(\mathbf{L}^{-1})_{23}=0$, the displacements δ_{I} and δ_{II} are not affected by the out-of-plane stress intensity factor K_{III}.

The normal displacement of the crack faces under mixed-mode loading is given by

$$\delta_{\mathrm{I},tip} = \sqrt{\frac{2r}{\pi}}[(\mathbf{L}^{-1})_{22}K_{\mathrm{I}} + (\mathbf{L}^{-1})_{12}K_{\mathrm{II}}] \tag{I2.1.3}$$

and for a mode-I loading exclusively,

$$\delta_{\mathrm{I},tip} = \sqrt{\frac{2r}{\pi}}(\mathbf{L}^{-1})_{22}K_{\mathrm{I}} = \sqrt{\frac{8r}{\pi}}\frac{K_{\mathrm{I}}}{E_{eff,22}} \tag{I2.1.4}$$

Vogelgesang [I2.2] gave the data for silicon nitride crystals, necessary for the computation of $\mathbf{L}^{-1}$ (for a graphic representation see [I2.3]). The effective module $E_{\mathrm{eff,22}}$ determining the displacements of (I2.1.4) by an Irwin parabola is shown in Fig. I2.2.

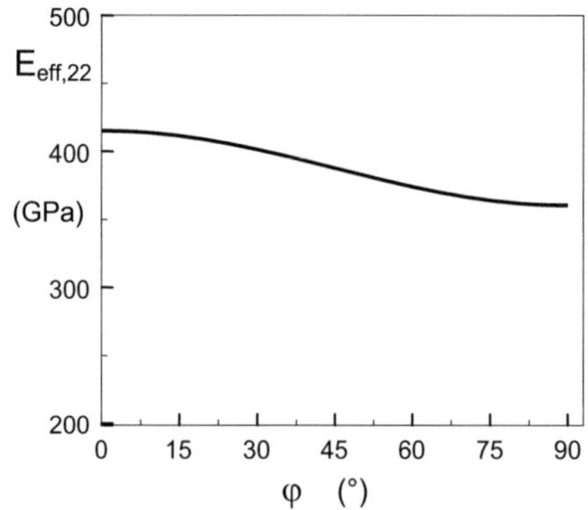

Fig. I2.2 Effective Young's modulus for application of eq.(I2.1.4).

I2.2 Experimental results

Figure I2.3 shows three different cracks extending in Vickers indentation tests with the tips arrested in a β-crystal (Fig. I2.3a to I2.3c) [I2.4]. In order to avoid a mode-II loading contribution, predominately cracks were selected that showed tangents at the crack tips which did not deviate too much from the direction of the crack path in the glass matrix before entering the crystal. Measurements of the total crack opening displacements δ_I for the 3 cracks terminating in the Si_3N_4 crystals under different crack orientation are represented in Fig. I2.4a.

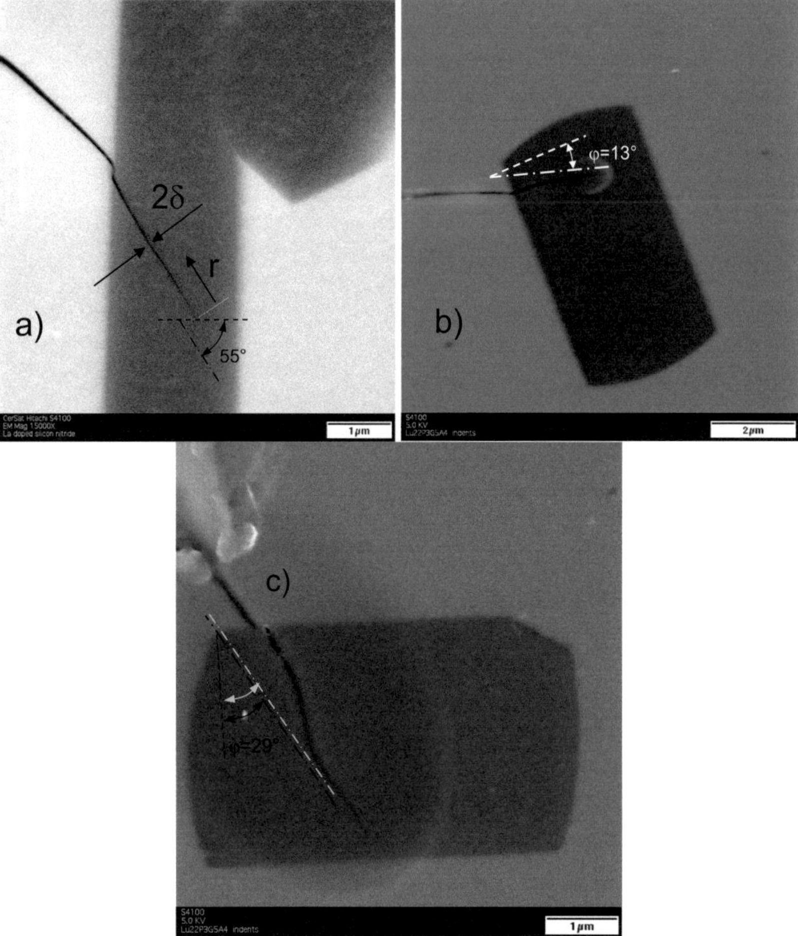

Fig. I2.3 a) Crack under 55° to the c-axis of a beta-crystal, b) crack-tip region under 13°, c) and under 29°.

The critical stress intensity factors for the β-crystal are introduced in Fig. I2.4b. From the tentatively introduced interpolation line we can expect that the maximum toughness of the β-crystal at φ=0 for crack propagation on the (0001)/basal plane is about 12

MPa√m. The minimum value for φ=90° can be expected at about 4 MPa√m for crack propagation on one of the prism planes. Trigonal structured sapphire (α-Al_2O_3) crystals is known to have a similar large anisotropy in fracture resistance [I2.5, I2.6].

Since the crack opening displacements of Fig. I2.4a show significant scatter, the error margin of the toughness data is admittedly large. It has to be taken into account that during indentation and further handling the indentation cracks could extend by subcritical crack growth in humid lab air with the consequence of a reduced actual stress intensity factor at the COD-measurements. From this point of view, the obtained results have to be interpreted as a **lower limit** for the toughness K_{Ic}.

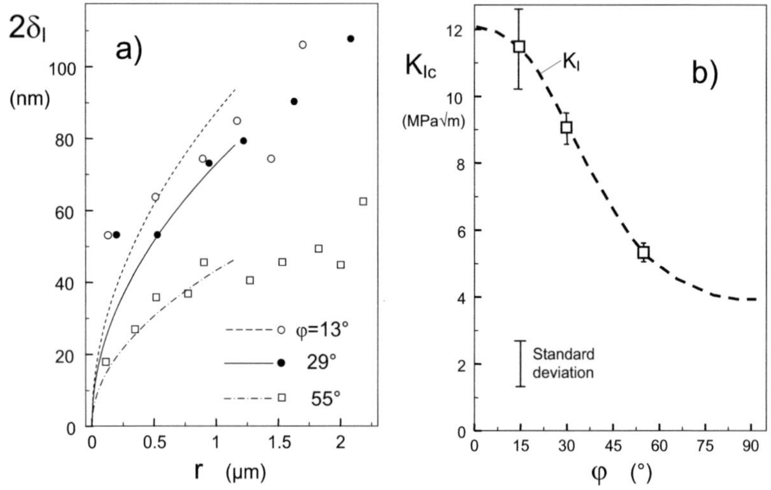

Fig. I2.4 a) Total crack opening for the cracks in Fig. I2.3 as a function of the crack-tip distance *r* with the fitting curves, b) fracture toughness data with tentatively introduced interpolation curve.

References I2

I2.1 Ting, T.C.T, Anisotropic elasticity: Theory and applications, Oxford University Press, 1996.

I2.2 Vogelgesang, R., Grimsditch, M., Wallace, J.S., The elastic constants of single crystal β-Si_3N_4, Appl. Phys. Letters **76**(2000), 982-84.

I2.3 Böhlke, T., Brüggemann, C., Graphical representation of the generalized Hook's law, Techn. Mech. **21**(2001), 145-158.

I2.4 S. Fünfschilling, M.J. Hoffmann, T. Fett, J. Wippler, T. Böhlke, P.F. Becher, unpublished work.

I2.5 S.M. Wiederhorn, Fracture of Sapphire, J. Am. Ceram. Soc. **52**(1969), 485-491.

I2.6 P. F. Becher, Fracture –Strength Anisotropy of Sapphire, J. Am. Ceram. Soc. **59**(1976), 59-61.

I3 Subcritical growth of Vickers indentation cracks

Measurement of crack opening displacements for Vickers indentation cracks in brittle materials is a rather simple method for the determination of the "crack-tip toughness" K_{I0}, the starting value of the R-curve (see Section D2). In such tests, it has to be ensured that subcritical crack growth in the time span between load removal and displacement evaluation under the scanning electron microscope (SEM) is negligible. The amount of crack extension can be computed if the parameters of the subcritical crack growth law are available for the investigated material.

During a Vickers indentation test in a ceramics surface, a cruciform semi-elliptical crack system is generated. Beneath the contact area of an indenter pressed into the surface, a residual stress zone remains even after unloading. Figure I3.1 gives the relevant geometric data of the crack system generally modelled by a single semi-elliptical surface crack. The parameter b is the radius of the damaged zone responsible for the residual stresses assumed semi-circular (Section E3). In realistic cases, see e.g. [I3.1] for Si_3N_4, the residual stress zone below the indenter is in the order of about $b/c \approx 0.5$.

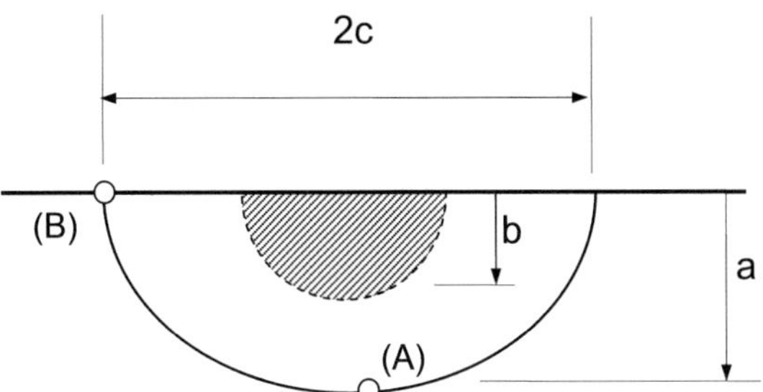

Fig. I3.1 Geometric data of an indentation crack of width $2c$ and depth a, opened by a residual stress zone of radius b.

A crack is considered, which is loaded in the centre region by a force P. This force is caused by the constant pressure p distributed over the area near the crack centre. For the stress intensity factors, it holds

$$K_{A,B} = \frac{2P}{(\pi a)^{3/2}} F_{A,B} = \frac{2P}{(\pi c)^{3/2}} \left(\frac{c}{a}\right)^{3/2} F_{A,B}$$

with

$$P = \tfrac{1}{2}\, p\pi\, b^2 \tag{13.1}$$

The condition $F_A = F_B$ resulting in $K_A = K_B$, is called the equilibrium condition. There is an influence of b/a on the equilibrium aspect ratio, roughly approximated for $b/a < 0.6$ by [13.2]

$$(a/c)_{eq} \cong 0.836 + 0.12\,(b/a)^2 \tag{13.2}$$

with the related geometric functions

$$F_A = F_B = 1.05 + 0.3315\,(b/a)^2 \tag{13.3}$$

Under stable crack-growth conditions at $K = K_{Ic}$, the conditions $K_A = K_B = K_{Ic}$ and, consequently, $a/c = (a/c)_{eq}$ are sufficiently fulfilled. Equations (13.2) and (13.3) read in terms of b/c (neglecting higher order terms than quadratic)

$$(a/c)_{eq} \cong 0.836 + 0.172\,(b/c)^2 \tag{13.2a}$$

$$F_B(c) \cong \lambda[1 + \mu(b/c)^2], \quad \lambda = 1.05,\ \mu = 0.45 \tag{13.3a}$$

Immediately after load removal it holds $K_A = K_B = K_{Ic}$ for $c = c_0$ and $a = a_0$. Due to subcritical crack growth, the crack dimensions a, c increase with time resulting in a decreasing stress intensity factor. The decreasing residual stress intensity factor at the surface point (B) reads for $c > c_0$

$$K_B(c) = K_{Ic}\left(\frac{c_0}{c}\right)^{3/2} \frac{F_B(c)}{F_B(c_0)} \tag{13.4}$$

Subcritical crack growth may be described by a power law relation

$$\frac{dc}{dt} = A^*\left(\frac{K_B}{K_{Ic}}\right)^n \tag{13.5}$$

For sufficiently steep da/dt-K-curves, e.g. $n > 10$, the crack develops under $a/c = (a/c)_{eq}$. Introducing (13.4) into (13.5) and using $F_B(c) = F_B(a/(a/c)_{eq})$, results in the differential equation

$$\left(\frac{c}{c_0}\right)^{3n/2}\left(\frac{F_B(c_0)}{F_B(c)}\right)^n dc = A^*\, dt \tag{13.6}$$

Taking the integral over dc from c_0 to c and over dt from 0 to t yields

$$\frac{c}{c_0}\, {}_2F_1[n,-\mu(\tfrac{b}{c})^2]^{\frac{2}{3n+2}} = \left({}_2F_1[n,-\mu(\tfrac{b}{c_0})^2]+\frac{3n+2}{2}\frac{A*}{c_0(1+\mu(\tfrac{b}{c_0})^2)^n}t \right)^{\frac{2}{3n+2}} \tag{I3.7}$$

with the hyper-geometric function ${}_2F_1$ in an abbreviated notation

$${}_2F_1[n,-x^2] \overset{def}{=} {}_2F_1[-\tfrac{1}{2}(1+\tfrac{3n}{2}),n,1+\tfrac{1}{2}(1-\tfrac{3n}{2}),-x^2] \tag{I3.8}$$

Equation (I3.7) is plotted in Fig. I3.2 for several values of b/c_0 and n. It can be seen that for identical n-values the slope of the $\lg(c)$-$\lg(t)$-curve varies with b/c_0.

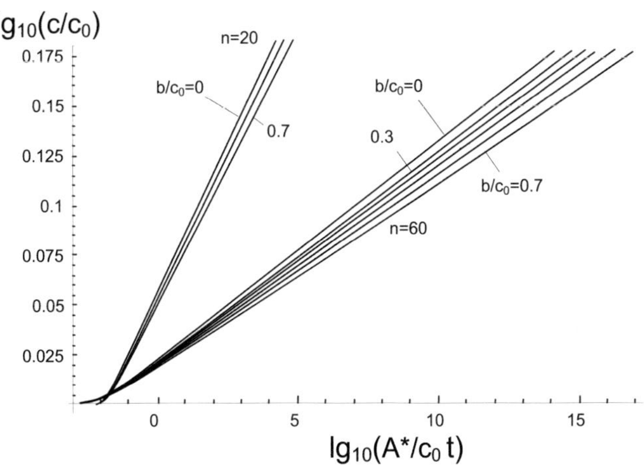

Fig. I3.2 Representation of eq.(I3.7).

For a simple evaluation of the rather unwieldy function ${}_2F_1$ one can apply the first two terms of the series expansion

$${}_2F_1[n,-x^2] \cong 1/(1+\tfrac{2n}{3n-2}x^2)^{\frac{3n+2}{2}} \tag{I3.9}$$

which represent the function in the interesting region of $x<0.35$ with an error less than 0.2%. With this approximation eq.(I3.7) gives

$$\frac{c}{1+\tfrac{2n}{3n-2}\mu(\tfrac{b}{c})^2} \cong c_0\left(\frac{1}{(1+\tfrac{2n}{3n-2}\mu(\tfrac{b}{c_0})^2)^{\frac{3n+2}{2}}} + \frac{3n+2}{2}\frac{A*}{c_0(1+\mu(\tfrac{b}{c_0})^2)^n}t \right)^{\frac{2}{3n+2}} \tag{I3.10}$$

 I3 Subcritical growth of Vickers indentation cracks

In the special case of a point-force approximation, $b\to0 \Rightarrow {}_2F_1\to1$, the well-known solution by Green and Sglavo [I3.3] results

$$\frac{c}{c_0} = \left(1 + \frac{3n+2}{2}\frac{A*}{c_0}t\right)^{\frac{2}{3n+2}}$$

(I3.11)

References I3

I3.1 Lube, T., Indentation crack profiles in silicon nitride, J. Europ. Ceram. Soc. **21**(2001), 211-218.
I3.2 Fett T. Stress intensity factors, T-stresses, Weight function (Supplement Volume), IKM55, KIT Scientific Publishing, Karlsruhe; 2009; (Open Access: available under http://digbib.ubka.uni-karlsruhe.de/volltexte/1000013835).
I3.3 Sglavo, M.V., Green, D.J., Subcritical crack growth of indentation median cracks in soda-lime-silica glass, J. A. Ceram. Soc. **78**(1995), 650-656.

I4 Diffusivity affected by swelling stresses

Diffusivities of glasses are commonly determined from the H-concentration profiles in the surface regions. In the context, it has to be considered that diffusion of water into silica surfaces is stress enhanced [I4.1]. Since water entrance will cause volume swelling [I4.2], the diffusion profiles become affected by the swelling stresses via

$$D = D_0 \exp(\Delta V_w \sigma_h / RT) \qquad (I4.1)$$

with the absolute temperature T, the gas constant R, the activation volume ΔV_w and the diffusivity D_0 at zero hydrostatic pressure σ_h.

In order to compute the diffusion profile, the partial diffusion differential equation for the uniaxial case with diffusivity depending on the water concentration

$$\frac{\partial C}{\partial t} = \frac{\partial}{\partial z}\left[D(C) \frac{\partial C}{\partial z} \right] \qquad (I4.2)$$

has to be solved with the hydrostatic stress proportional to the water concentration $C=C_w$ according to eqs.(H3.1.7) and (H3.1.9). This was numerically done by use of the line method in the Mathematica routine NDSolve [I4.3] with the result plotted in Fig. I4.1a. In this representation the abbreviation

$$\alpha C_0 = \frac{\Delta V_w \sigma_{h,0}}{RT} \, , \qquad (I4.3)$$

is used with the water content at the surface, C_0, and the hydrostatic swelling stress $\sigma_{h,0}$ resulting from swelling at the surface ($z=0$).

Figure I4.1b again shows the numerically computed profile for $\alpha C_0 \approx -1$ as the continuous curve. It seems reasonable to fit the same numerical data with the diffusion profile for stress-free diffusion with an apparent diffusivity D_{app}:

$$\left(\frac{C}{C_0} \right)_{num} = \mathrm{erfc}\left(\frac{z}{2\sqrt{D_{app} t}} \right) , \qquad (I4.4)$$

Different fitting procedures for the determination of D_{app} are possible. In the following the apparent diffusivity may be obtained from the condition of same integrals

I4 Diffusivity affected by swelling stresses

$$\int_0^\infty \left(\frac{C}{C_0}\right)_{num} dz = \int_0^\infty \mathrm{erfc}\left(\frac{z}{2\sqrt{D_{app}\, t}}\right) dz \tag{I4.5}$$

The effective erfc-function obtained from the condition (I4.5) is shown in Fig. I4.1b by the dash-dotted curve.

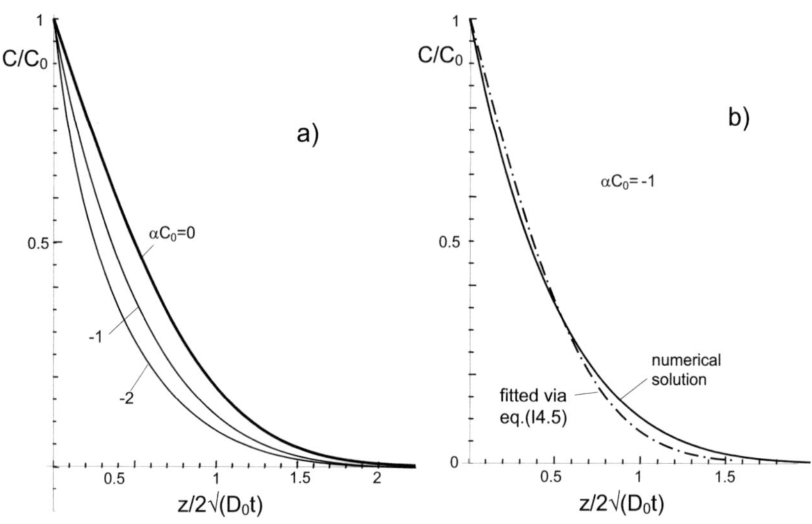

Fig. I4.1 a) Diffusion profiles for (negative) swelling stresses, b) numerical result fitted by the commonly used erfc-profiles.

In Fig. I4.2 the apparent diffusivity D_{app} is plotted as a function of αC_0. The curve can be approximated by

$$\frac{D_{app}}{D_0} \cong \exp[0.656(\alpha C_0) + 0.04(\alpha C_0)^2] \tag{I4.6}$$

This relation allows the true diffusivity D_0 to be evaluated from measured diffusivities D_{app}.

In the case of silica at 88°C, the volumetric swelling strain obtained from the results of Shelby [I4.2] and Zouine et al. [I4.4] is $\varepsilon_v \cong 0.325\%$ [I4.5]. This results in $\sigma_{h,0}=-63.5$ MPa and with $\Delta V_w=15$ cm^3/mol (Section H3.4) according to eq.(I4.3) in $\alpha C_0 \approx -0.318$. In this case eq.(I4.6) yields $D_0=1.216\, D_{app}$.

Let us now consider the hydrogen diffusion profile measured at 100°C by Zouine et al. [I4.4]. At this temperature, the swelling strain is slightly increased to $\varepsilon_v \cong 0.364\%$ as can be concluded from the surface water concentration by Zouine et al. [I4.4] shown in Fig. I4.3a, roughly described for the range of 0°C-200°C by

$$c_0(H) \cong 0.894 \times 10^{22} / cm^3 \exp\left(-\frac{Q}{RT}\right) \tag{I4.7}$$

(Q=10.54 kJ/mol). Consequently it holds $\sigma_{h,0} \cong -71$ MPa and $\alpha C_0 \approx -0.344$. From (I4.6) we get now $D_0 \cong 1.236\, D_{app}$.

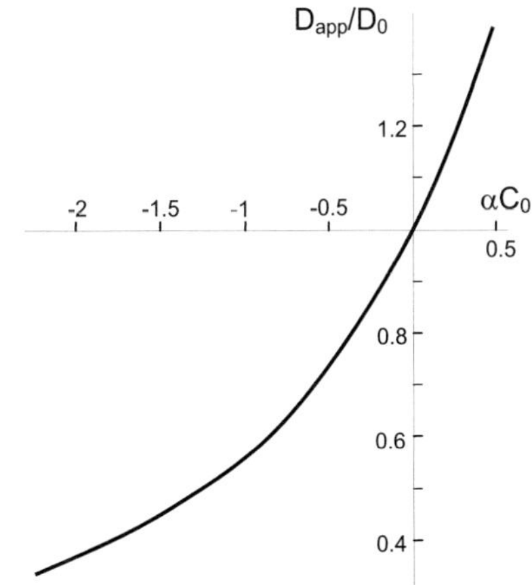

Fig. I4.2 Apparent diffusivity D_{app} as a function of αC_0.

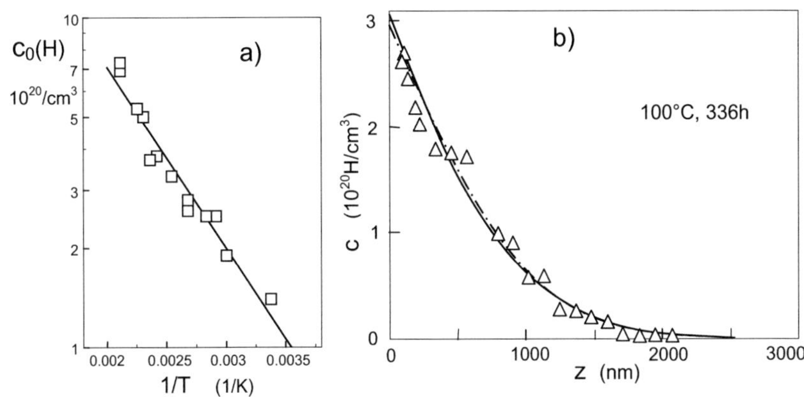

Fig. I4.3 a) Surface (H)-concentration in silica for liquid water by Zouine et al.[I4.4], b) hydrogen diffusion profile in silica after 336 h at 100°C, also measured by Zouine et al. (triangles) compared with the fit according to eq.(I4.5) (dash-dotted curve) and the result of numerical evaluation of eq.(I4.2) (continuous curve).

 I4 Diffusivity affected by swelling stresses

Zouine et al. [I4.4] fitted the measured profile for t=336h with a diffusivity of D_{app}=2.9 10^{-19} m^2/s, from which the stress-free diffusivity D_0=3.57 10^{-19} m^2/s results.

Figure I4.3b shows the experimental data (triangles) and the fit according to eq.(I4.4) as the dash-dotted curve. The solid curve represents the numerical evaluation of eq.(I4.2). A difference can hardly be detected, i.e. within the experimental scatter both curves fit the same data set. This makes it nearly impossible to determine ε_v or ΔV from diffusion profiles at rather low temperatures.

References I4

I4.1 M. Nogami and M. Tomozawa, Effect of stress on water diffusion in silica glass, *J. Am. Ceram. Soc.*, **67** 151-154 (1984).

I4.2 Shelby, J.E., Density of vitreous silica, J. Non-Cryst. **349**(2004), 331-336.

I4.3 *Mathematica*, Wolfram Research, Champaign, USA.

I4.4 Zouine, A., Dersch, O., Walter, G., Rauch, F., Diffusivity and solubility of water in silica glass in the temperature range 23-200°C, Phys. Chem. Glasses, **48** (2007), 85-91.

I4.5 T. Fett, G. Rizzi, M.J. Hoffmann, S. M. Wiederhorn, Effect of Water on the inert Strength of Silica Glass: Role of Water Penetration, to be submitted .

I5 Modifications of shielding by additional effects

I5.1 Deformation of steady-state swelling zones

The computations of the swelling zones in Sections H3.3 and H3.4 were carried out for diffusion in a motionless coordinate system with a fixed crack tip, i.e. for an *arrested* crack. Whereas for a *growing* crack the transversal diffusion normal to the crack plane is hardly affected by the moving crack tip, the diffusion and the crack propagation compete in crack direction [I5.1, I5.2]. The consequence for strongly stress enhanced diffusion zones with cordial shape is a reduced ratio of the zone size in length direction r_0 to the zone height ω, which for an arrested crack is [I5.3]

$$\frac{r_0(0)}{\omega(0)} = \frac{8}{\sqrt{27}} \tag{I5.1.1}$$

($r_0(0)$, $\omega(0)$ stands for $r_0(v{=}0)$ and $\omega(v{=}0)$, respectively).

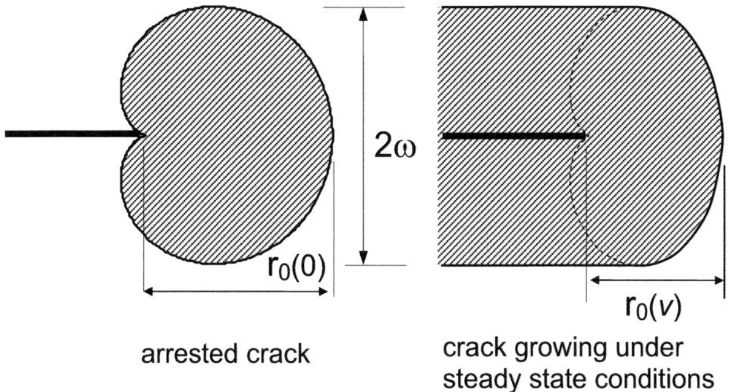

Fig. I5.1 a) Swelling zone ahead the tip of an arrested crack, b) expected reduced zone length for a crack growing at a constant crack-growth rate v (schematic).

The effect of a moving crack tip on the zone shape is schematically shown in Fig. I5.1. The ratio $r_0(v)/r_0(0)$, computed according to [I5.1] and [I5.2] is plotted in Fig. I5.2 versus the logarithm of the crack rate v. The arrows indicate some crack-growth rates. For crack rates of $v{>}10^{-12}$ m/s as relevant for subcritical crack growth experiments it was found that $0.75{<}r_0(v)/r_0(0){<}1$.

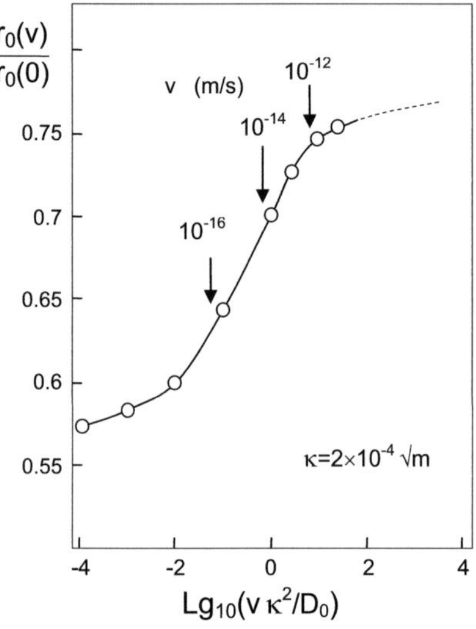

Fig. I5.2 Deformation of the swelling zone, (for κ see eq.(H3.2.7)), arrows indicate some subcritical crack growth rates.

If we assume that the diffusion normal to the crack-plane direction is not affected by the crack rate since there exists no normal velocity component ($v_\perp=0$), we obtain for subcritical crack growth rates $v>10^{-12}$ m/s

$$\tfrac{3}{4}\frac{8}{\sqrt{27}}<\frac{r_0(v)}{\omega(v)}<\frac{8}{\sqrt{27}}\qquad(15.1.2)$$

For the stress intensity factor of the deformed zones one can use some single results from literature.

The stress intensity factors for three different ratios of r_0/ω are available from [I5.3]. These cases are illustrated in Fig. I5.3a. Figure I5.3b shows the related coefficients ψ for the general representation

$$K_{sh}=-\psi\,\frac{\varepsilon_0 E}{1-\nu}\sqrt{\omega}\qquad(15.1.3)$$

by the solid circles (for the symbols in eq.(I5.1.3) see Section H3). The interpolating curve through the data points can be expressed simply by

$$\psi\cong0.37\left(1-\tfrac{1}{2}\tanh\!\left(\frac{7}{9}\frac{r_0}{\omega}\right)\right)\qquad(15.1.4)$$

For the lower value in eq.(I5.1.2), $r_0/\omega \cong 2/\sqrt{3}$, $\psi=0.238$ is obtained as indicated in Fig. I5.3b by the open circle.

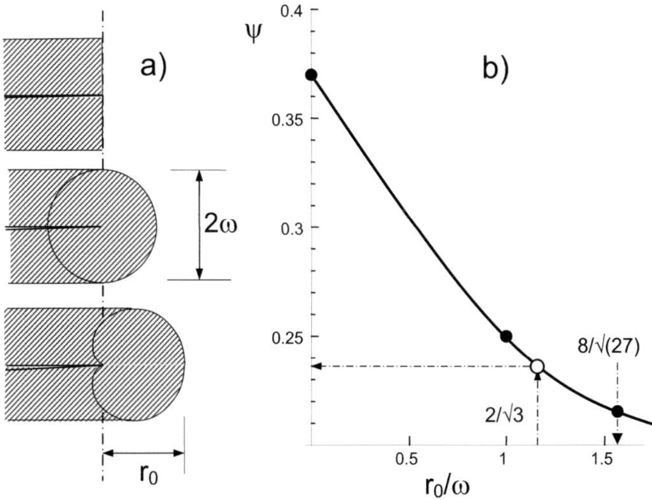

Fig. I5.3 Shielding stress intensity factors by eq.(I5.1.3) for a deformed zone with reduced length ahead the crack.

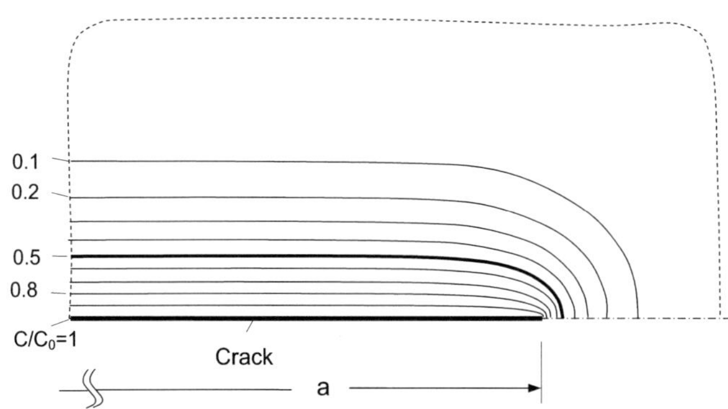

Fig. I5.4 Diffusion of water through crack surfaces of a semi-infinite crack, contour lines for constant water concentration C normalized on the crack-surface concentration C_0 [I5.4].

In a FE-study by Rizzi [I5.4], the diffusion problem for a water-soaked *semi-infinite edge crack in an infinite body* was solved. The water concentration profiles are shown in Fig. I5.4. It can be seen that the zone size ahead of the tip is reduced.

213

In contrast to the assumptions made in Fig. I5.3, there is also a height reduction of diffusion zones visible near the tip. From the stress intensity factor evaluation by ABAQUS 6.2 a shielding coefficient of $\psi=0.255$ was obtained.

I5.2 Crack leaving a swelling zone

At higher temperatures and after long time, also unloaded cracks may develop a swelling zone of thickness b by water diffusion into the crack faces (Fig. I5.5a.). If the swelling strain is constant in the swelling zone, $\varepsilon_v=\varepsilon_0=$const., and $\varepsilon_v=0$ outside the zone, the shielding stress intensity factor is

$$K_{sh} = -\psi\,\frac{\varepsilon_0 E}{1-v}\sqrt{b} \qquad (15.2.1)$$

with the coefficient ψ depending on the zone shape in the crack-tip region. In order to minimize the uncertainty in the true shape of the zone ends it was recommended in [15.5] to use for computations $r_0/b = 0.5$ with the related coefficient $\psi = 0.3$.

If such a crack grows by an amount of Δa, it will escape from the initial swelling zone, as illustrated in Fig. I5.5b-I5.5d for different zone ends. The shielding stress intensity factors for two limit cases of the zone shape near the crack tip are plotted in Fig. I5.5e. For the circular zone end (Fig. I5.5b), the shielding stress intensity factor is given by the squares [15.7]. A straight zone end with $r_0/b=0$ yields the dash-dotted curve. The two dependencies can be described by eq.(I5.2.1) where now ψ is a function of $\Delta a/b$.

The stress intensity factor solution for the case $r_0/b=\frac{1}{2}$ (Fig. I5.5d) can be obtained by linear interpolation of the two limit cases resulting in the dashed curve of Fig. I5.5e. The coefficient ψ may be approximated by

$$\psi \cong \begin{cases} 0.305 + \frac{1}{12}\Delta a/b & for \quad \Delta a/b \le \frac{1}{2} \\ 0.34\exp[-2\sqrt{\Delta a/b - \frac{1}{2}}\ ] & \Delta a/b > \frac{1}{2} \end{cases} \qquad (15.2.2)$$

It should be mentioned that the curves in Fig. I5.5e change abruptly caused by the assumption of a sharp layer boundary. For an arbitrarily varying strain distribution, the weight function or Green's function method from [15.6] can be used.

The stress intensity factor caused by a swelling layer of thickness db' in the surface distance $z=b'$ (Fig. I5.6a) undergoing the strain $\varepsilon(b')$ results as

$$d(K_{sh}) = K_{sh}(b'+db') - K_{sh}(b') \qquad (15.2.3)$$

This gives rise for a Green's- or "weight function" h defined as

$$h = \frac{d(K_{sh})}{db'}$$

(I5.2.4)

For any swelling distribution $\varepsilon(z)=\varepsilon(b')$, the related shielding stress intensity factor is then obtained by summing over all the zone increments db' multiplied with the local swelling strain

$$K_{sh} = \frac{1}{\varepsilon_0} \int_0^\infty \varepsilon(b')\,h(b')\,db', \quad \varepsilon_0 = \varepsilon(b'=0)$$

(I5.2.5)

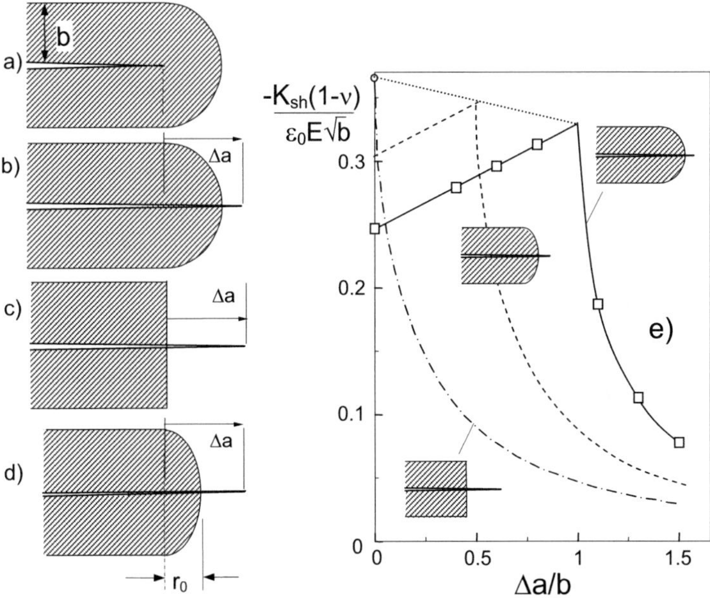

Fig. I5.5 a) Crack with a thick circular swelling layer of constant swelling strain, b) crack after additional extension Δa for a circular zone, c) for a rectangular zone, d) for a zone with $r_0/b=1/2$, e) variation of the shielding stress intensity factor for a crack growing in and outside the initial layer according to [I5.3] (dash-dotted curve) and [I5.7] (squares); dashed curve: interpolated from the continuous and dash-dotted curves for $r_0/b=1/2$.

In the absence of stress-enhanced diffusion, the swelling profiles are represented by the complementary error function. The volumetric strain at the location $z=b'$ is in this case

$$\varepsilon = \varepsilon_0 \operatorname{erfc}\left(\frac{b'}{2\sqrt{D_0\,t}}\right)$$

(I5.2.6)

where ε_0 is the strain at the free surface, D_0 the diffusion coefficient and t the time after the first water contact.

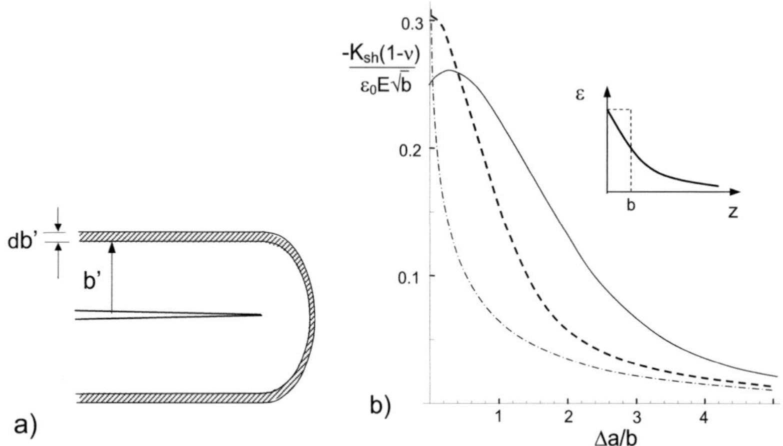

Fig. I5.6 a) Green's function as the shielding stress intensity factor of a thin swelling layer of thickness b', b) stress intensity factors according to the dashed curve in (d) for an erfc-shaped strain distribution (insert); curves according to Fig. I5.5e.

Equation (I5.2.5) has been evaluated for the case of $r_0/b=\frac{1}{2}$. The resulting shielding stress intensity factors related to the erfc-shaped swelling profiles are shown in Fig. I5.6b. The numerical result for $r_0/b=\frac{1}{2}$ can approximated by equation (I5.2.1) with

$$\psi \approx \frac{0.305}{1+\left(\dfrac{\Delta a}{b}\right)^2} \tag{I5.2.7}$$

I5.3 Decrease of Young's modulus

Due to the water content in glasses, a reduction of the Young's modulus was observed. This effect has been reported by Ito and Tomazawa [I5.8] for a Na_2SiO_2 glass. Figure I5.7 shows the Young's modulus of the water containing glass E_w, normalized on the modulus of the dry bulk material, versus the volumetric swelling strain ε_v. The modulus at zero water content ($C_w=0$) may be denoted as $E^{(0)}$. From the interpolation curve for the data points a reduction of 10% is observable at about $\varepsilon_v =17\%$. The initially linear dependency reads roughly

$$\frac{E_w}{E^{(0)}} \approx 1 - \tfrac{1}{2}\varepsilon_v \qquad (15.3.1)$$

as indicated in Fig. 15.7 by the dashed straight line.

An effect similar that for the Na_2SiO_2 glass might also be expected for silica. Such a module reduction must affect the shielding stress intensity factor because K_{sh} is proportional to the product of strain ε and Young's modulus E

$$K_{sh} \propto \varepsilon E \qquad (15.3.2)$$

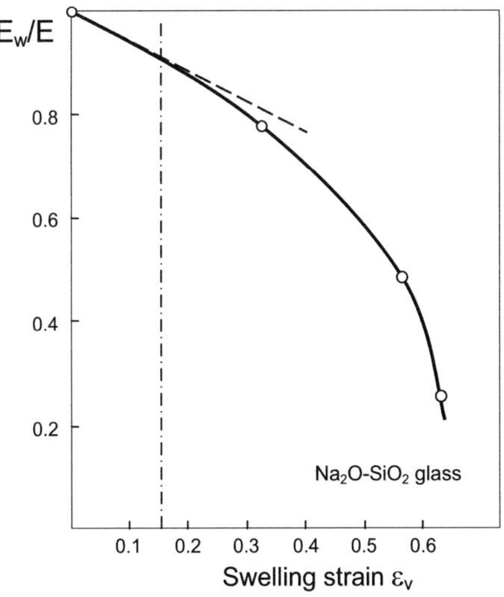

Fig. 15.7 Decrease of Young's modulus due to water content as a function of volumetric swelling strain for a Na_2SiO_2 glass by Ito and Tomazawa [15.8].

The effect of water on the volume strain reads in series expansion

$$\varepsilon = \varepsilon^{(0)} + \frac{d\varepsilon}{dC_w}C_w + O(C_w^2) = \frac{d\varepsilon}{dC_w}C_w + O(C_w^2) \qquad (15.3.3)$$

with the strain $\varepsilon^{(0)}=0$ for $C_w=0$. The effect of ε is of **first** order in C_w. For the Young's modulus it holds

$$E = E^{(0)} + \frac{dE}{dC_w}C_w + O(C_w^2) \qquad (15.3.4)$$

217

The Young's modulus of course has a non-disappearing value (the maximum) at $C_w=0$, i.e. $E^{(0)}\neq0$.

The product $\varepsilon\times\mathbf{E}$ must then also be of first order in C_w

$$K_{sh} \propto \varepsilon\,E \to E^{(0)}\frac{d\varepsilon}{dC_w}C_w + O(C_w^2) \tag{15.3.5}$$

A remarkable influence of the modulus may be present for very high water concentrations that can make a modified K_{sh}-computation necessary.

For the most general case of modulus and strain both depending of the water content, and consequently on the crack-tip distance, $E(r)$ and $\varepsilon(r)$, an effective product $(E\times\sqrt{\varepsilon})_{eff}$ can be defined by the Green's function procedure of [15.6]. For this purpose the varying modulus has to be drawn under the integral sign of eq.(H3.4.1) resulting in

$$(E\sqrt{\omega})_{eff} = \frac{1}{2\varepsilon_0}\int_0^\infty \varepsilon\,E\,h(\omega')\,d\omega' \tag{15.3.6}$$

Next, a direct influence of a reduced Young's modulus on the applied stress intensity factor must be taken into account. Evans and Faber [15.9] have treated this general problem very early. They considered the special case of the modulus reduced by formation of a micro-cracking zone around a crack tip. For any time-independent material behavior, the fracture mechanics J-integral by Rice [15.10] can be used as the loading parameter. It simply reads for linear-elastic materials

$$J = \frac{K^2(1-v^2)}{E} \tag{15.3.7}$$

and is identical to the energy release rate G.

Since the J-integral for any path around the crack tip is independent of the specially chosen path, its value must be the same for a path Γ_2 far away from the tip (in the bulk) and the path Γ_1 directly at the crack tip, i.e.

$$\frac{K_{appl,2}^2(1-v^2)}{E} = \frac{K_{appl,1}^2(1-v_1^2)}{E_1} \tag{15.3.8}$$

where E_1 and v_1 are the elastic properties at the tip affected by water. Consequently, the applied stress intensity factor at the crack tip is

$$K_{appl,1} = K_{appl,2}\sqrt{\frac{E_1}{E}\frac{1-v^2}{1-v_1^2}} \approx K_{appl,2}\sqrt{\frac{E_1}{E}} \tag{15.3.9}$$

Since a reduced module in the crack-tip region reduces the effective applied stress intensity factor, the modulus effect may also be counted on the shielding side as a *formally computed* additional shielding term ΔK_{sh}

$$\Delta K_{sh} = K_{appl,1} - K_{appl,2} \approx K_{appl,2}\left(\sqrt{\frac{E_1}{E}} - 1\right) \qquad (I5.3.10)$$

This instantaneous shielding term is not necessarily a constant value defined by the externally applied loads exclusively. It must depend on humidity and temperature. From eq.(I5.3.10) it is obvious that this type of a shielding term is independent on the height of the water diffusion zone as well as on the crack-growth increment Δa.

As emphasized in [I5.9] the effect of shielding by micro-cracking may be counteracted by a reduced crack resistance of the damaged material. So far, it is not sufficiently known how strong the silica is "damaged" by the water diffusion. It is also unknown how strong the crack-growth resistance is reduced. Consequently, we cannot conclude what the net-effect is, i.e. whether the reduced crack-tip loading or the reduced material resistance prevail.

References I5

I5.1 Zhao, T., Nehorai, A., Detection and localization of a moving biochemical source in a semi-infinite medium, Report, Electrical and Computer Engineering Dept. Univ. of Illinois, Chicago, USA.

I5.2 Babich, V.M., On asymptotic solutions of the problems of moving sources of diffusion and oscillations, UDC 534.231.1, 517.226, 1991, pp. 3067-3071, Plenum Publishing Corp.

I5.3 McMeeking, R.M., Evans, A.G., Mechanics of Transformation-Toughening in Brittle Materials, J. Am. Ceram. Soc. **65**(1982), 242-246.

I5.4 Rizzi, G., unpublished work (2012).

I5.5 S.M. Wiederhorn, T. Fett, G. Rizzi, M. Hoffmann, J.-P. Guin, Water Penetration – its Effect on the Strength and Toughness of Silica Glass, Krakow Stress Corrosion Symposium, August, 2011, submitted to *Metallurgical and Materials Transactions*

I5.6 S. M. Wiederhorn, T. Fett, G. Rizzi, M.J. Hoffmann, J.-P. Guin, The Effect of Water Penetration on Crack Growth in Silica Glass, to appear in Engng. Fract. Mech.

I5.7 Fett, T., Guin, J.P., Wiederhorn, S.M., Interpretation of effects at the static fatigue limit of soda-lime-silicate glass, Engng. Fract. Mech. **72**(2005), 2774-279.

I5.8 Ito, S., Tomozawa, M., J. de Phys. **C9**(1982), 611.

I5.9 Evans, A.G., Faber, K.T., Crack-growth resistance of microcracking brittle materials, J. Am. Ceram. Soc. **67**(1984), 255-260

I5.10 Rice, J.R., A path independent integral and the approximate analysis of strain concentration by notches and cracks, Trans. ASME, J. Appl. Mech. (1986), 379-386.

I6 Passive R-curve testing device

I6.1 Load-displacement behaviour

I6.1.1 Displacement-controlled bending

Several different systems are known in literature, which ensure stable crack-growth tests on bending bars. Here, only two systems may be mentioned:

- A purely passive load control was reached in [I6.1, I6.2] by using a very stiff spring element parallel to the specimen as was originally proposed by Markowski [I6.3] for tensile test devices (Maniette et al. [I6.4] used the principle for CT-specimens).

- Test devices with combined passive and active elements were used by Moon et al.[I6.5] (for a photographic representation see e.g. [I6.6]) and by Jelitto et al. [I6.7]. The passive elements were also rather rigid frames and piezo actuators guaranteed a fast load reduction at the onset of first indication for instable crack extension.

The R-curve experiments of Section 2.2 were performed with the testing device described in detail in [I6.7]. In this section, the purely mechanical device for stable crack growth tests may be described.

For controlled fracture tests, a test arrangement with very low compliance is necessary. The device shown in Fig. I6.1 (proposed in [I6.1]) fulfils this condition. This system shows nearly disappearing machine compliance as will be explained in the following considerations. For this purpose, the compliances of the device are replaced in Fig. I6.1b by concentrated compliances. The compliance of the very stiff frame is C_{FR} and the compliance of the load cell C_{LC}. Other vagabonding compliances in the central load path as for instance the elastic deformation of the loading and supporting rollers may be included in C_{LC}. C_{SP} represents the compliance of the pre-cracked test specimen.

Since the load path through the specimen and the load cell act parallel to the frame, the related displacements must be identical

$$\delta_{SP} + \delta_{LC} = P_{SP}(C_{SP} + C_{LC}) = \delta_{FR} = P_{FR}C_{FR} \tag{16.1.1}$$

where P_{SP} is the load applied to specimen and load cell and P_{FR} the load carried by the frame. The total load P is given as

$$P = P_{SP} + P_{FR} \qquad (I6.1.2)$$

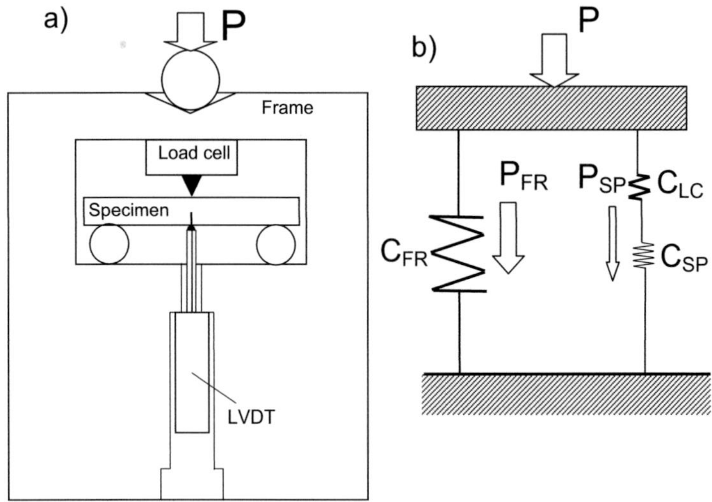

Fig. I6.1 a) Testing device ensuring stable crack extension, b) compliance model with concentrated compliances symbolized by elastic springs.

Combining eqs.(I6.1.1) and (I6.1.2) yields for the load on the test specimen

$$P_{SP} = P \frac{C_{FR}}{C_{FR} + C_{SP} + C_{LC}} \qquad (I6.1.3)$$

and the specimen displacement

$$\delta_{SP} = P_{SP} C_{SP} = P \frac{C_{FR}}{C_{FR} + C_{SP} + C_{LC}} C_{SP} \qquad (I6.1.4)$$

Let us now assume that at a certain fixed load P a crack may virtually grow unstably by an infinitesimal increment da. The load on the specimen drops immediately by an amount of dP_{SP}

$$\frac{dP_{SP}}{da} = -P \frac{C_{FR}}{(C_{FR} + C_{SP} + C_{LC})^2} \frac{dC_{SP}}{da} = -\frac{P_{SP}}{C_{FR} + C_{SP} + C_{LC}} \frac{dC_{SP}}{da} \qquad (I6.1.5)$$

Simultaneously the displacement of the bar changes by

$$\frac{d\delta_{SP}}{da} = P\frac{C_{FR}(C_{FR}+C_{LC})}{C_{FR}+C_{SP}+C_{LC}}\frac{dC_{SP}}{da} \qquad (16.1.6)$$

If the frame compliance and the load-cell compliance are small compared to that of the bending bar, *i.e.*, if

$$C_{FR}, C_{LC} \ll C_{SP} , \qquad (16.1.7)$$

eqs.(I6.1.3) to (I6.1.6) simplify to

$$P_{SP} = P\frac{C_{FR}}{C_{SP}} \qquad (16.1.8a)$$

or equivalently $\qquad \Rightarrow \dfrac{dP_{SP}}{da} = -\dfrac{PC_{FR}}{C_{SP}^{\,2}}\dfrac{dC_{SP}}{da} = -\dfrac{P_{SP}}{C_{SP}}\dfrac{dC_{SP}}{da} \qquad (16.1.8b)$

and $\qquad\qquad \delta_{SP} = P_{SP}C_{SP} = PC_{FR} = const, \qquad (16.1.9a)$

i.e. $\qquad\qquad\qquad \Rightarrow \dfrac{d\delta_{SP}}{da} \cong 0 \qquad (16.1.9b)$

As the consequence of condition (I6.1.7), the specimen displacements become independent of the specimen compliance, i.e. they would remain constant for an assumed onset of spontaneous crack extension. The load instantaneously decreases.
The condition (I6.1.7) can be reached by use of a very rigid frame, a quartz load cell and a rather deeply notched specimen.

I6.1.2 Stability of crack extension

The stabilization discussed before considers an externally reached displacement controlled fracture. However, even in the limit case of condition (I6.1.7) a crack can become unstable under special conditions which may be denoted here as an "internal stability".
This stability of crack growth depends on the equilibrium of elastically stored energy in the bending bar and the energy release rate necessary for crack extension. It can be influenced by an appropriate choice of specimen- and supporting dimensions and on the initial notch depth. Many papers in literature on ceramics deal with this topic (e.g. [16.8], [16.9]).
A short derivation of the basic equations may be given according to [16.10]. The energy for crack extension is provided by two sources, the work A done by the external

forces and the elastically stored energy in the component U. The energy release rate G_I in case of a virtual increase of the crack by the unit area is

$$G_\mathrm{I} = \frac{\mathrm{d}A}{\mathrm{d}S} - \frac{\mathrm{d}U}{\mathrm{d}S} \ , \tag{16.1.10}$$

where $S=B\,a$ is the crack area (B specimen thickness).

After a crack extension Δa the compliance has increased. The load decreases by an amount of ΔP_SP. The work $\mathrm{d}A$ done by the externally applied forces is zero, because the total load point displacement has been kept constant, eq.(16.1.9b).

In the following, the energy release rate is computed for the general case, where both the displacement and the load may change during crack extension.

The elastically stored energy is

$$U = \frac{1}{2}P_{SP}\delta_{SP} \tag{16.1.11}$$

and the energy release rate according to eq.(16.1.10)

$$G = \frac{1}{2}P_{SP}^2 \frac{\mathrm{d}C_{SP}}{B\mathrm{d}a} \ . \tag{16.1.12}$$

This can be expressed in terms of stress intensity factors by using the Irwin relation:

$$K_\mathrm{I}^2 = G_\mathrm{I}E' \tag{16.1.13}$$

with

$$E' = \begin{cases} E & \text{for plane stress} \\ E/(1-v^2) & \text{for plane strain} \end{cases} \ . \tag{16.1.14}$$

Equation (16.1.12) then reads

$$K_\mathrm{I} = P_{SP}\sqrt{\frac{1}{2}E'\frac{\mathrm{d}C_{SP}}{B\mathrm{d}a}}\Bigg|_{\delta=const} = \frac{\delta_{SP}}{C_{SP}}\sqrt{\frac{1}{2}E'\frac{\mathrm{d}C_{SP}}{B\mathrm{d}a}} \tag{16.1.15}$$

In 4-point bending, it holds according to eq.(B1.5.2)

$$C_{SP} = \frac{9}{2}\frac{L^2\pi}{BW^4E'}\int_0^a a'F^2(a')\mathrm{d}(a') + C_0 \tag{16.1.16}$$

with the supporting span L in 3-point bending or $L=S_1\text{-}S_2$ in 4-point bending ($S_1=$ supporting roller span, $S_2=$ loading roller span), the geometric function for stress intensity factors, F, and the compliance C_0 of the bar without a crack. Further geometric data are the width W and the thickness B of the bar.

The derivative occurring in eq.(I6.1.15) results as

$$\frac{dC_{SP}}{da} = \frac{9}{2}\frac{(S_1 - S_2)^2}{E'BW^4}\,\pi\,aF^2$$

(I6.1.17)

The compliance C_0 is known from bending theory as

$$C_0 = \left(\frac{S_1 - S_2}{W}\right)^2 \frac{1}{E'B}\left[\frac{S_1 + 2S_2}{4W} + \frac{(1+\nu)W}{2(S_1 + S_2)}\right]$$

(I6.1.18)

where the second term in brackets is small compared to the first one.

Now it is assumed that a crack with initial size a_0 is grown to an increased total crack length $a_1 = a_0 + \Delta a_1$. For this crack, the related applied stress intensity factor $K_I(a_1)$ must equal the crack growth resistance, *i.e.*,

$$K_I(a_1) = K_R(\Delta a_1)$$

(I6.1.19)

The stress intensity factor for a virtually increased length $a = a_1 + da$ can then be written as

$$K_I(a) = \underbrace{\frac{C(a_1)}{C(a)}\sqrt{\frac{a}{a_1}}\,\frac{F(a)}{F(a_1)}}_{\lambda}K_R(\Delta a_1)$$

(I6.1.20)

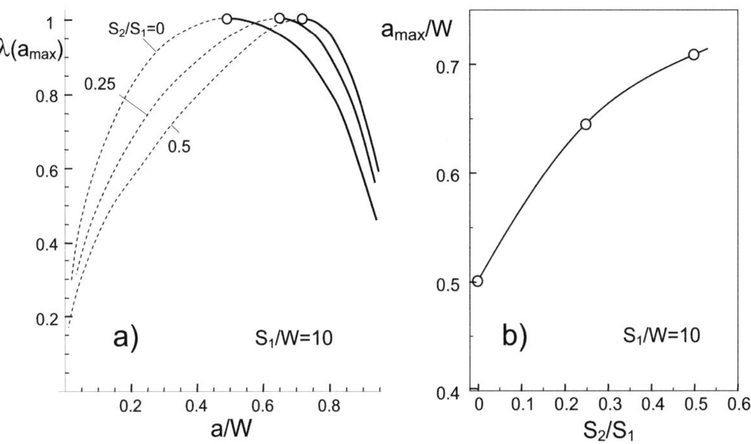

Fig. I6.2 Crack-growth stability: a) Coefficient λ of eq.(I6.1.20) as a function of relative crack depth a/W and inner loading roller span S_2, b) effect of S_2/S_1 on the relative crack depth for maximum of λ.

Equation (I6.1.20) can be used for stability computations. Stability is guaranteed as long as the condition

$$\frac{dK_1(a)}{da} \leq \frac{dK_R(\Delta a)}{da}$$

(I6.1.21)

is fulfilled. This condition reads for the most serious case, a material without any rising R-curve, $dK_R(\Delta a)/da=0$. In this case, a disappearing or negative $dK_1(a)/da$ is necessary to allow stable crack growth.

The coefficient λ in eq.(I6.1.20), normalized on its maximum value at a_{max}, is plotted in Fig. I6.2a for different ratios of the inner and the outer roller span as a function of a/W. The solid curve parts indicate the ranges of crack stability. The dashed curve parts represent the regions of positive $dK_1(a)/da$ in which stable crack extension is impossible so far the material shows a flat K_R-curve. Figure I6.2b gives the minimum relative crack length as a function of the ratio S_2/S_1. It can be concluded that a stable crack growth test in 3-point bending ($S_2=0$) can be performed at lower relative crack depths a/W and allows K_R to be determined in a larger region Δa than in a standard 4-point bending test with $S_2/S_1=$ ½. In the case of silicon nitrides with very steep initial $K_R(\Delta a)$-curves, stability is ensured also for cracks with $a < a_{max}$.

The crack length for maximum stress intensity factors is plotted in Fig. I6.3 as a function of the ratios S_2/S_1 and S_1/W. In order to ensure an extended region of stability these two ratios should be chosen as small as possible. The lower limit of the two parameters is caused by the applicability of stress intensity factor equations from literature valid mostly for a limited S_1/W range.

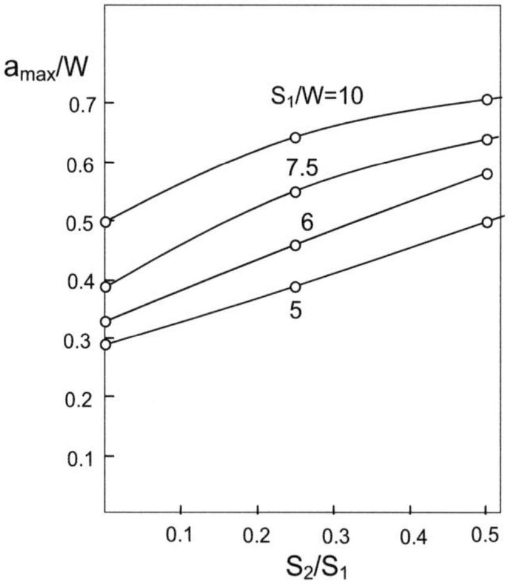

Fig. I6.3 Effect of S_2/S_1 and S_1/W on the relative crack depth for maximum stress intensity factors.

16.1.3 Stable and unstable crack extension

From the curves in Fig. 16.2a also the possibility of crack arrest after a spontaneous crack extension phase can be predicted. Since the curves for large a/W show strongly negative slopes, eq.(16.1.21) can always be fulfilled. If for instance a crack in a material with a flat crack resistance curve exhibits an initial length $a/W<0.5$, even in 3-point bending the crack must extend unstable but *may* stop when the curve slope is sufficiently negative. Unfortunately, this crack arrest cannot be guaranteed from eq.(16.1.21) since this relation represents a quasi-*static* failure condition. The reason is the excess energy, which is accompanied with the fast crack extension similar to the crack arrest situation in thermal shock considerations.

In Fig. 16.4 the occurrence of stable and unstable crack propagation is shown for a 3-point bending test with $S_1/W=10$ and cracks with different initial depths in a material with constant crack-growth resistance. Since crack arrest must be discussed in energy terms, the ordinates are scaled in energy release rates, *i.e.*, in K^2.

In Fig. 16.4a a crack depth of $a/W=0.2$ is chosen. Under monotonically increasing load (dashed curves) the condition $K_I=K_{Ic}$ is reached for the solid curve. Spontaneous crack extension starts. From the static equilibrium condition, we would conclude that this crack should stop at about $a/W\approx0.75$ since then K_I falls below K_{Ic}. From energy balance considerations, it is obvious that the crack at this location has an energy excess represented by the hatched area W_1 above the K_{Ic}-line. This energy is available for further unstable crack extension. Crack arrest will occur, when the area W_2 below the K_{Ic}-curve fulfils $W_1=W_2$. This is the case at about $a/W\approx1$, representing the limit for completely unstable tests.

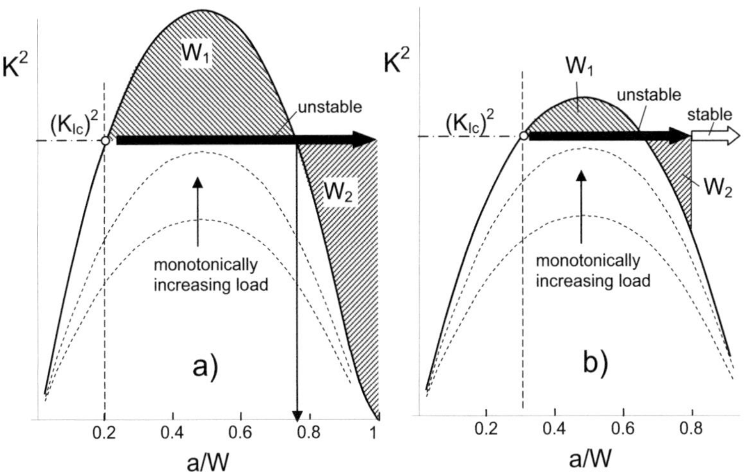

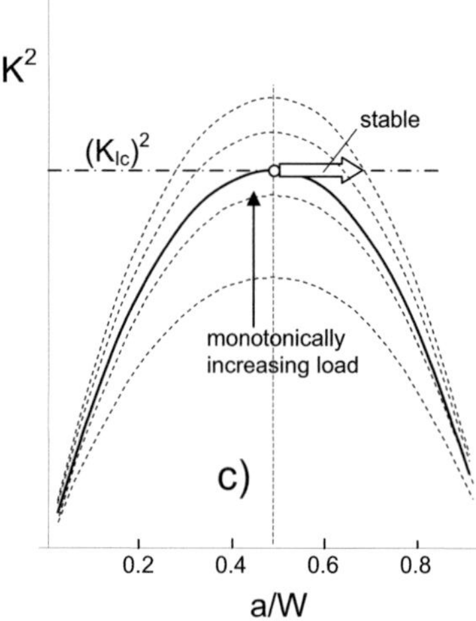

Fig. I6.4 Propagation of cracks with different initial depth, for 3-point bending and S_1/W=10.

The case of a larger crack with a/W=0.3 is shown in Fig. I6.4b. For this crack also unstable crack extension occurs since at the onset of crack extension dK_I/da>0. It is obvious that the excess energy is now clearly reduced. Consequently, also the additional unstable crack increment is shorter.

Finally, Fig. I6.4c illustrates the extension of a crack with initial depth a/W=0.5. This crack cannot show any instability. With increasing load, the arrow for the crack length extension always ends at the intersection of K_I and K_{Ic}. Static equilibrium is always fulfilled. The same holds of course for any crack with initial depth a/W>0.5.

The crack extension behaviour in displacement-controlled R-curve tests is very close to that occurring in thermal shock tests [I6.11, I6.12]. Literature on thermal shock also covers the effect of a rising R-curve [I6.13], [I6.14] (see also Chapter 11 in [I6.10]).

I6.1.4 Determination of displacements

The compliance used for the computation of the actual crack depth is defined as the displacement at the inner roller contact relative to that at the outer rollers. In order to avoid uncertainties due to flattening of the rollers, displacement measurements should be performed directly on the specimen. Since the rollers prevent the direct application of displacement pickups at the contact line, the displacements may be determined at the centre of the specimen and at a location near the supporting rollers (with distance β from the contact line) as illustrated in Fig. I6.5. The difference of the two LVDT gives a displacement δ_1 that is of course different from the desired displacement δ. Following [16.15], the compliance $C=\delta/P$ can be computed in very good approximation from the measured one $C_1=\delta_1/P$ via

$$C = \kappa C_1 + \chi C_0 \tag{16.1.22}$$

with C_0 from eq.(16.1.18) and the abbreviations

$$\kappa = \frac{S_1 - S_2}{S_1 - 2\beta} \, , \tag{16.1.23}$$

$$\chi = 1 - \kappa\left(1 + \frac{\frac{3}{4}cS_2^2 - 3\beta c(c + S_2) + \beta^3}{2c^3 + 3S_2 c^2}\right), \tag{16.1.24}$$

$$c = \frac{S_1 - S_2}{2} \tag{16.1.25}$$

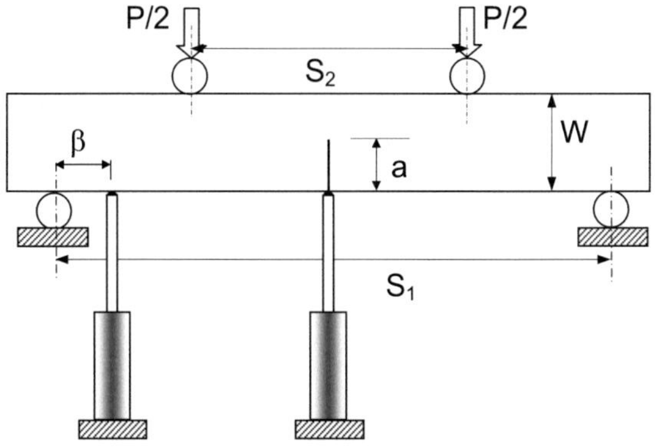

Fig. I6.5 Measurement of the bending displacements according to [16.15].

16.2 Experimental results

The authors of [16.16] used the system sketched in Fig. I6.1a with the 3-point bending arrangement replaced by 4-point load application. A pre-notched bending bar of width $W=4$mm, thickness $B=3.5$mm and notch depth $a_0>2$mm was loaded in bending. In the load path, a quartz cell recorded the load acting on the specimen. Two displacement pick-ups measured the bending displacements directly at the specimen from which the displacements at the centre with respect to the line contacts between rollers and specimen could be computed as proposed in [16.15]. In this way, supporting effects and

Hertzian deformation of the supporting rollers could be eliminated from the experimental data.

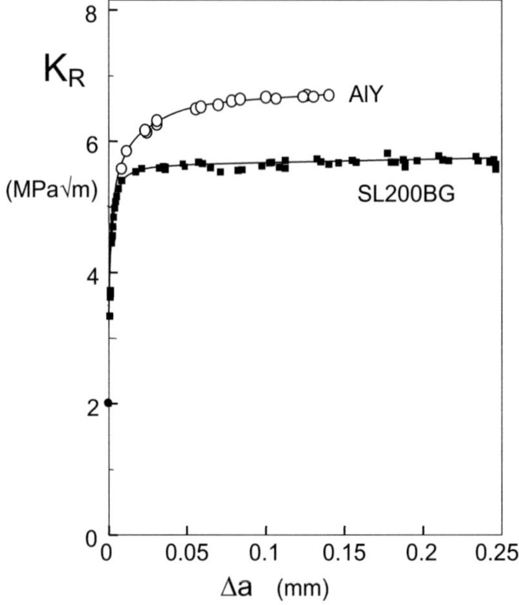

Fig. I6.6 R-curves for two silicon nitrides with 3wt% Y_2O_3 and 3wt% Al_2O_3 content.

As an example of application, R-curve measurements were carried out on a silicon nitride with 3wt% Y_2O_3 and 3wt% Al_2O_3 content (denoted here as AlY) produced by Sumitomo (Japan). This material shows a composition very similar to the SL200 (CeramTec, Plochingen) studied in [I6.17] (see also Section 2.2). Wagner et al. [I6.16] carried out strength, toughness, and R-curve measurements. The characteristic strength in 4-point bending was found to be σ_0=1405 MPa with a Weibull module of m=10.3. Fracture toughness measurements on SEVNB-specimens with notch-root radii <7μm resulted in K_{Ic}= 6.5±0.26 MPa√m.

The R-curve for material AlY is shown in Fig. I6.6 by the circles. In addition, the SL200 data from Fig. C2.2 are introduced for comparison. Similar to the fracture toughness, also the R-curve is by the same percentage (≈15-20 %) above that of SL200 (K_{Ic}= 5.65±0.25 MPa√m).

References I6

I6.1 T.Fett, D. Munz, G. Thun, H.A. Bahr, Evaluation of bridging parameters in Al2O3 from R-curves by use of the fracture mechanical weight function, J. Amer. Ceram. Soc. **78**(1995) 949-51.
I6.2 G. Rauchs, T. Fett, D. Munz, R-curve behaviour of 9Ce-TZP zirconia ceramics, Engng. Fract. Mech. **69**(2002), 389-401.
I6.3 W. Markowski, A new principle of material testing machines (in German), Materialprüfung **32**(1990), 144-48.
I6.4 Y. Maniette, M. Inagaki, M. Sakai, Fracture toughness and crack bridging of a silicon nitride ceramic, J. Eur. Ceram. Soc. **7**(1991), 255-63.
I6.5 Moon, R., Bowman, K., Trumble, K., Rödel, J., Comparison of R-curves from single-edge V-notched-beam (SEVNB) and surface-crack-in-flexure (SCF) fracture toughness tests methods on multilayered alumina-zirconia composites, J. Am. Ceram. Soc. **83**(2000),445-447.
I6.6 T. Fett, D. Munz, A.B. Kounga Njiwa, J. Rödel, G.D. Quinn, Bridging stresses in sintered reaction-bonded Si3N4 from COD measurements, J. Europ. Ceram. Soc. **25**(2005), 29-36.
I6.7 H. Jelitto, F. Felten, M. V. Swain, H. Balke, G. A. Schneider: Measurement of the Total Energy Release Rate for Cracks in PZT under Combined Mechanical and Electrical Loading. J. Appl. Mech. **74**(2007) 1197 – 1211.
I6.8 Kleinlein, W., Slow crack growth in brittle materials carried out in bending tests (in German), PhD-thesis, University Erlangen-Nürnberg, 1980.
I6.9 C. M. Peret, J. A. Rodrigues, Stability of crack propagation during bending tests on brittle materials, Cerâmica **54**(2008), 382-387.
I6.10 Munz, D., Fett, T., *CERAMICS*, Failure, Material Selection, Design, Springer-Verlag (1999).
I6.11 Swain, M.V. (1990): R-curve behaviour and thermal shock resistance of ceramics, J. Am. Ceram. Soc. **73**, 621–628.
I6.12 Swain, M.V. (1993): Significance of non-linear stress-strain and R-curve behaviour on thermal shock of ceramics, in: Thermal Shock and Thermal Fatigue Behaviour of Advanced Ceramics, Eds. G.A. Schneider and G. Petzow, Kluwer Academic Publishers, Dordrecht, Netherlands.
I6.13 Bahr, H.A., Fett, T., Hahn, I., Munz, D., Pflugbeil, I. (1993): Fracture mechanics treatment of thermal shock and the effect of bridging stresses, in: Thermal Shock and Thermal Fatigue Behaviour of Advanced Ceramics, Eds. G.A. Schneider and G. Petzow, Kluwer Academic Publishers, Dordrecht, Netherlands.
I6.14 Schneider, G.A., Magerl, F., Hahn, I., Petzow, G. (1993): In situ observations of unstable and stable crack propagation and R-curve behaviour in thermally loaded disks, in: Thermal Shock and Thermal Fatigue Behaviour of Advanced Ceramics, Eds. G.A. Schneider and G. Petzow, Kluwer Academic Publishers, Dordrecht, Netherlands.
I6.15 Fett, T., Diegele, E., Indirect measurements of compliances in four-point-bending tests, J. of Testing and Evaluation, **16**(1988), 487-88.
I6.16 Wagner, S., Fünfschilling, S., Fett, T., Creek, D., Hoffmann, M. J., R-curves for silicon nitride determined with a passive R-curve device, unpublished work, KIT-IAM (2011), Karlsruhe.
I6.17 Fünfschilling, S; Fett, T; Oberacker, R; Hoffmann, MJ; Oezcoban, H; Jelitto, H; Schneider, GA; Kruzic, JJ, R-curves from compliance and optical crack-length measurements, J. Am. Ceram. Soc., **93**(2010), 2814-21.

Schriftenreihe
des Instituts für Angewandte Materialien

ISSN 2192-9963

Die Bände sind unter www.ksp.kit.edu als PDF frei verfügbar oder
als Druckausgabe bestellbar.

Band 1   Prachai Norajitra
          Divertor Development for a Future Fusion Power Plant. 2011
          ISBN 978-3-86644-738-7

Band 2   Jürgen Prokop
          **Entwicklung von Spritzgießsonderverfahren zur Herstellung von
          Mikrobauteilen durch galvanische Replikation.** 2011
          ISBN 978-3-86644-755-4

Band 3   Theo Fett
          **New contributions to R-curves and bridging stresses – Applications
          of weight functions.** 2012
          ISBN 978-3-86644-836-0